BEHAVIORAL SYNTHESIS AND COMPONENT REUSE WITH VHDL

BEHAVIORAL SYNTHESIS AND COMPONENT REUSE WITH VHDL

Ahmed A. JERRAYA
Hong DING
Polen KISSION
Maher RAHMOUNI
TIMA/INPG
Grenoble, France

CONTRIBUTION BY

E. BERREBI, W. CESARIO, P. GUILLAUME
TIMA/INPG
Grenoble, France

KLUWER ACADEMIC PUBLISHERS
Boston/London/Dordrecht

Distributors for North America:
Kluwer Academic Publishers
101 Philip Drive
Assinippi Park
Norwell, Massachusetts 02061 USA

Distributors for all other countries:
Kluwer Academic Publishers Group
Distribution Centre
Post Office Box 322
3300 AH Dordrecht, THE NETHERLANDS

Library of Congress Cataloging-in-Publication Data

A C.I.P. Catalogue record for this book is available
from the Library of Congress.

Copyright © 1997 by Kluwer Academic Publishers

All rights reserved. No part of this publication may be reproduced, stored in a retrieval system or transmitted in any form or by any means, mechanical, photo-copying, recording, or otherwise, without the prior written permission of the publisher, Kluwer Academic Publishers, 101 Philip Drive, Assinippi Park, Norwell, Massachusetts 02061

Printed on acid-free paper.

Printed in the United States of America

CONTENTS

LIST OF FIGURES — ix

LIST OF TABLES — xv

PREFACE — xvii

1 **INTRODUCTION** — 1
  1.1 System Design: the Productivity Bottleneck — 1
  1.2 From Physical Design to System Design: Abstraction Levels — 2
  1.3 Behavioral Synthesis — 5
  1.4 Design Reuse — 17
  1.5 Component Reuse in VLSI — 18
  1.6 Modular Design Methodology For Component Reuse at the Behavioral Level — 20
  1.7 Summary — 22

2 **MODELS FOR BEHAVIORAL SYNTHESIS** — 23
  2.1 Design representation for behavioral synthesis — 23
  2.2 The datapath controller model — 33
  2.3 Datapath models — 40
  2.4 Controller models — 58
  2.5 Summary — 64

3 **VHDL MODELING FOR BEHAVIORAL SYNTHESIS** — 67
  3.1 Interpretation of VHDL descriptions — 68
  3.2 Behavioral VHDL execution modes — 90

	3.3	Scheduling VHDL descriptions	94
	3.4	Summary	124

4 BEHAVIORAL VHDL DESCRIPTION STYLES FOR DESIGN REUSE — 127

	4.1	Design reuse	127
	4.2	Design reuse at the behavioral level	135
	4.3	Modular design	141
	4.4	VHDL modeling for reuse	144
	4.5	Towards object oriented design in VHDL	150
	4.6	Summary	154

5 ANATOMY OF A BEHAVIORAL SYNTHESIS SYSTEM BASED ON VHDL — 155

	5.1	Main principles	156
	5.2	Design steps and execution models	162
	5.3	Interactive synthesis	185
	5.4	Behavioral synthesis in the design loop	194
	5.5	Summary	200

6 CASE STUDY: HIERARCHICAL DESIGN USING BEHAVIORAL SYNTHESIS — 203

	6.1	The PID	203
	6.2	Specifications	204
	6.3	System-level analysis and partitioning	205
	6.4	Hierarchical Design	205
	6.5	Design for reuse of the fixed-point unit as a behavioral component	206
	6.6	Abstraction for reuse	209
	6.7	Design reuse	209
	6.8	The behavioral synthesis process	213
	6.9	Summary	215

7 CASE STUDY: MODULAR DESIGN USING BEHAVIORAL SYNTHESIS — 217

	7.1	Introduction	217

7.2	System specification	218
7.3	System partitioning	221
7.4	Behavioral specifications of subsystems	227
7.5	System design	240
7.6	Behavioral and RTL simulations	247
7.7	Summary	248

REFERENCES 249

INDEX 259

LIST OF FIGURES

Chapter 1

1.1	A Multi-level Description System (Answering Machine)	5
1.2	System Design Process	6
1.3	A typical data flow oriented application	8
1.4	A typical DSP application: mixing data flow oriented and control flow oriented designs	9
1.5	Behavioral synthesis steps	14
1.6	Use of Modular Design Methodology	21

Chapter 2

2.1	Example of a simple data flow graph	25
2.2	(a) Algorithmic description of GCD. (b) CFG of the GCD example	26
2.3	Combined control-data flow graph for GCD example.	28
2.4	Separate control-data flow graph for GCD example.	29
2.5	(a)FSMD Model, (b) FSMD implementation	30
2.6	FSMC Architecture Model	31
2.7	(a)FSMC Model, (b) FSMC implementation	32
2.8	General Architecture Model	34
2.9	Schedule of events inside a circuit.	36
2.10	A simple clocking scheme.	38
2.11	A clocking scheme with a controller pipelining.	39
2.12	Introduction of an Idle cycle.	40
2.13	Conceptual organization of the data-path architecture	41
2.14	Generic model of a datapath unit	42
2.15	Functional unit representation	43
2.16	Example of component pipelining	44
2.17	Basic SU set	45

2.18	Basic communication units structure	46
2.19	Example of communication through switches and MUXs	48
2.20	Basic set of external communication units	49
2.21	General transfer	50
2.22	Example of particular transfer : chaining	51
2.23	Example of concurrent operations	53
2.24	Datapath pipelining scheme	54
2.25	Transfers execution for a bus-based architecture	55
2.26	Bus model: synthesis example	56
2.27	Transfers executions for a mux-based architecture	57
2.28	Multiplexer model: synthesis result	57
2.29	General Synchronous Sequential System Organization	59
2.30	Sequencing of the gcd example	61
2.31	Moore Automaton : VHDL Description	62
2.32	Mealy Automaton : VHDL Description	62
2.33	Micro-programmable controller architecture	63

Chapter 3

3.1	Excerpt VHDL model of a Bubble sort	69
3.2	Complex system specification example	70
3.3	Interpretation of conditional constructs	72
3.4	Writing styles for a conditional construct	73
3.5	Process synchronization	75
3.6	Telephone answering machine; (a) block diagram of system; (b) StateChartlike model of controller.	76
3.7	VHDL representation of the controller; (a) Loop hierarchy; (b) Code segment.	77
3.8	Loop execution, (a) sequential, (b) unrolling, (c) folding	78
3.9	VHDL description example	82
3.10	In-line expansion of the procedure	83
3.11	Procedure as an independent module	84
3.12	Procedure as a coprocessor	85
3.13	(a) Original signal assignments, (b) Redundant signal assignment	86

List of Figures   xi

3.14 Mapping variables to registers using the left-edge algorithm, (a) description example, (b) sorted variable lifetime intervals, (c) resulting registers   87
3.15 Array to Memory Mapping   89
3.16 High-level FSM representation of a behavioral VHDL description   90
3.17 (a) Data-flow scheduling. (b) Control-flow scheduling.   94
3.18 Incomplete Data Flow based VHDL process description   97
3.19 Data flow graph and ASAP schedule   97
3.20 List schedule with resource constraint of one multiplier and one adder   98
3.21 (a) probability of scheduling operations into control steps, (b) initial addition distribution graph   100
3.22 (a) probability of scheduling operations into control steps after operation 5 is assigned to control step 2, (b) new addition distribution graph   101
3.23 ILP scheduling as a TRCS   104
3.24 VHDL description of *ab mod n* function   105
3.25 CFG of the *ab mod n* function.   106
3.26 Scheduling one path using AFAP scheduling   110
3.27 (a)Scheduled paths. (b) FSM for AFAP scheduling   111
3.28 Path generation algorithm for DLS   112
3.29 Paths and successors for the Dynamic Loop Scheduling   113
3.30 FSM for the Dynamic Loop Scheduling   114
3.31 Scheduling one path using PPS   115
3.32 (a) Scheduled Paths for PPS, (b)FSM for PPS.   116
3.33 CDFG model of a conditional branch   119
3.34 Control-flow graph, VHDL process description   120
3.35 Paths and FSM schedule example   121
3.36 Performance analysis based scheduling result   122
3.37 DFG and Schedule of Conditions   122

Chapter 4

4.1 Design reuse during behavioral synthesis   127
4.2 Design reuse   128
4.3 Use of structured design methodology   131

4.4	Modular architecture	132
4.5	Reuse of the behavioral description	134
4.6	The four abstraction levels	136
4.7	Using communication primitives	138
4.8	Design partitioning	140
4.9	Control system with PID	140
4.10	Partitioning of the initial PID description	142
4.11	Modeling functional unit with local memory for reuse	144
4.11	Architecture behavior A	144
4.11	Modeling functional unit with local inputs and outputs for reuse	145
4.11	Architecture behavior B	145
4.12	Pure VHDL modeling for reuse	146
4.13	Description Representation	147
4.14	Pure behavioral architecture	148
4.15	Mixed behavioral-structural architecture	149
4.16	FSMC model	149
4.17	Pid description	151

Chapter 5

5.1	Synthesis by incremental refinements	154
5.2	Incremental refinement model	156
5.3	Behavioral description structure	157
5.4	Target architecture	158
5.5	AMICAL design flow	159
5.6	Synthesis flow	161
5.7	Examples of behavioral descriptions	164
5.8	The base components on the library	166
5.9	AMICAL technology file example	167
5.10	The four abstraction levels	167
5.11	Organization of the behavioral description	170
5.12	Control flow graph models	171
5.13	Development state after the scheduling: Behavioral FSMC	172
5.14	Behavioral finite state machine models	173
5.15	Example of link between operations and resources	175
5.16	Behavioral finite state machine with resources model	175

List of Figures

5.17	Micro-scheduling model for multi-cycle operations	177
5.18	RTL finite state machine model	178
5.19	Authorized transfers	178
5.20	Bus-based solution for the GCD example	179
5.21	Mux-based solution for the GCD example	180
5.22	Virtual architecture	182
5.23	Re-programmable controller architecture	183
5.24	AMICAL at work	185
5.25	Design flow for interactive synthesis	187
5.26	AMICAL at work	189
5.27	Evaluation and statistics reports	190
5.28	Incremental refinement based behavioral synthesis	191
5.29	Design loops.	193
5.30	Mixing synthesizable and nonsynthesizable parts	196
5.31	Different models used for the validation of the design and the design process	198

Chapter 6

6.1	PID algorithm (VHDL extract)	203
6.2	Hierarchical PID design	204
6.3	Full synthesis process (from behavior to gate)	205
6.4	Fixed-point unit description	206
6.5	Fixed-point unit interface	207
6.6	2-step protocol for multiplication	208
6.7	PID algorithm described in VHDL	211
6.8	Synthesis results of the fixed-point unit datapath with AMICAL	212
6.8	Synthesis results of PID datapath with AMICAL	212

Chapter 7

7.1	A global architecture of a Window Searching algorithm based System	217
7.2	Visual representation of the search process	218
7.3	Search algorithms	218
7.4	Abstract architecture of the *WSS*	221
7.5	Modular use of high-level synthesis	221

7.6	Package with subtype declarations	223
7.7	Component declaration package	225
7.8	Connections of the *co_processor*	226
7.9	Abstract architecture of the *co_processor*	227
7.10	Activation protocol between the top controller and the coprocessor	228
7.11	Behavioral description of the coprocessor	230
7.12	Datapath generation of *co_processor* by AMICAL	230
7.13	Connection of the *mem_sequence*	231
7.14	Abstract architecture of the *mem_sequence*	231
7.15	Writing in the RAM	232
7.16	behavioral description of *mem_sequence*	234
7.17	Datapath generation of *mem_sequence* by AMICAL	234
7.18	Behavioral description of *mem_string*	236
7.19	Datapath generation of *mem_string* by AMICAL	237
7.20	Connections of the WSS	237
7.21	Activation protocol of *mem_string*	238
7.22	VHDL behavioral description of the top controller instanced within a structural architecture of the WSS	240
7.23	VHDL structural description of the *WSS*	241
7.24	VHDL behavioral description of the *WSS*	243
7.25	Datapath generation of *WSS* by AMICAL	244
7.26	Mixed level simulations	245

LIST OF TABLES

Chapter 1

1.1 Timing Concepts and Specification Levels through the Design Process   3

Chapter 3

3.1 Interpretation modes of behavioral VHDL   93
3.2 Results for the *ab mod n* algorithm.   117

Chapter 5

5.1 Architectural models   162
5.2 Design tasks in design loops   194

PREFACE

Complex integrated circuits design for the 2000's is facing an increasing design productivity gap: over the past years, design productivity has increased by 21% per year on an average, while technology capability was increasing by 61% per year. This gap is becoming a limiting factor for the exploitation of the capabilities of the coming technologies. It is then clear that we need improvements of the design quality and designers' productivity. This may be achieved in two ways that can be combined:

- Using more structured design methodologies for an extensive reuse of existing components and subsystems. It seems that 70% of new designs correspond to existing components that cannot be reused because of a lack of methodologies and tools.

- Providing higher level design tools allowing to start from a higher level of abstraction. After the success and the widespread acceptance of logic and RTL synthesis, the next step is behavioral synthesis, commonly called architectural or high-level synthesis.

This book provides methods and techniques for VHDL based behavioral synthesis and component reuse. The goal is to develop VHDL modeling strategies for emerging behavioral synthesis tools. A special attention is given to structured and modular design methods allowing hierarchical behavioral specification and design reuse. The goal of this book is not to discuss behavioral synthesis in general or to discuss a specific tool but to provide the specific issues related to behavioral synthesis of VHDL description.

This book targets designers who have to use behavioral synthesis tools or who wish to discover the real possibilities of this emerging technology. The reader should have previous knowledge of logic and RTL design. Some knowledge of VHDL is also required. The book also targets teachers and students interested to learn or to teach VHDL based behavioral synthesis.

Chapter 1 discusses current design methodologies and outlines the need for improvements of the design quality and designers' productivity through higher level design tools such as behavioral synthesis and extensive reuse of existing components and subsystems. Classical behavioral synthesis techniques are also introduced. Finally the chapter discusses the requirement for behavioral synthesis and design reuse based on VHDL.

Chapter 2 includes some basic informations needed for behavioral synthesis. It discusses the main intermediate format and the main architectural models used for behavioral synthesis. The concepts introduced in this chapter will be used in the subsequent chapters.

Chapter 3 discusses the VHDL constructs from a behavioral synthesis point of view. Several writing styles will be discussed. The goal is to show the hardware semantics implied by each writing style. Several scheduling techniques will be presented.

Chapter 4 introduces VHDL modeling strategies that handle behavioral hierarchy and design reuse. The main models used are those of the component and the system. A component model is a subsystem that will be reused. A system is a whole design made of an assembling of already specified (or designed) components. These two concepts will be detailed and used for structuring behavioral VHDL descriptions. The basic idea behind this modeling strategy is that a complex system is generally composed of a set of subsystems performing specific tasks. A high-level specification of such a system needs only to describe the sequencing of these tasks, consequently the coordination of the different subsystems. Each subsystem is modeled as a functional module designed (or selected) to perform a set of specific operations. Therefore the behavioral specification may be seen as a coordination of the activities of the different subsystems. The decomposition of a system specification into a global control and detailed tasks allows to handle very complex design through hierarchy.

Chapter 5 deals with a specific behavioral synthesis tool called AMICAL. AMICAL combines behavioral synthesis with methodologies allowing design reuse. The goal is to detail the different design models and synthesis steps used by a behavioral synthesis tool in order to show the correspondence between the behavioral VHDL description and the produced architecture. Several of the models and techniques described in this chapter are common to several VHDL behavioral synthesis tools.

Chapter 6 analyzes hierarchical design. The design is decomposed hierarchically into a set of subsystems or components. Behavioral synthesis will be used

Preface

for the design of the subsystems which will be used in a second step as behavioral components for the behavioral description of the overall system. Finally behavioral synthesis is applied to the full system.

Chapter 7 deals with modular design. A large system is decomposed into a set of modules. Each module is synthesized independently using behavioral synthesis. The key issue in this kind of design is to use a modeling style that allows the different subsystems to be simulated at different levels of abstraction.

Acknowledgments

We would like to take this opportunity to thank many people who have helped in making this book possible:

All the sponsors of our work in high level synthesis

- Special thanks to Joseph Borel from SGS-Thomson who is supporting heartedly behavioral synthesis and design reuse.
- SGS-Thomson, the JESSI and ESPRIT programs for their generous funding of our work.
- All those who contributed with their industrial experience and their recommendations to make behavioral synthesis practical; in particular, we wish to mention Jean-Pierre Moreau, Jean Fréhel, Jean-Pierre Schoellkopf, Michel Harrand, Jean-Claude Herluison and Pierre Paulin from SGS-Thomson, Jacques Lecourvoisier and Etienne Closse from France Telecom/CNET and Laurent Bergher from TCEC.

All those who made the hard work of building the behavioral synthesis tool AMICAL: Kevin O'Brien, Inhag Park, Vijay Raghavan and Richard Pistorius.

A.A. Jerraya, H. Ding, P. Kission, M. Rahmouni

&

E. Berrebi, V. Cesario, P. Guillaume

Grenoble, France

BEHAVIORAL SYNTHESIS AND COMPONENT REUSE WITH VHDL

1
INTRODUCTION

1.1 SYSTEM DESIGN: THE PRODUCTIVITY BOTTLENECK

Nowadays one of the major objectives within VLSI domain is the improvement of the design quality and of the designers' productivity. This is due to the fact that the design process is characterized by 2 sets of factors: constant factors and variable ones. For instance typical large design budgets are usually fixed to around 10 to 15 persons over 18 months, and designers' productivity has been evaluated to some 10 objects per day. Controversely, ASIC design complexity has been increasing exponentially since 1984 from a hundred thousand transistors, to reach 1 to 2 million transistors today. The design complexity forecast for the year 2000 is 7 million transistors for a 0.18 micron CMOS technology. It is expected that this exponential increase will continue until year 2010 in order to reach 40 million transistors [SIA94].

At present, when using the most advanced logic and RTL (register transfer level) synthesis tools, such a budget covers the design of only 1 to 2 million transistors ASICs. This means that for year 2010 we still have a factor of 20 to 40 to close the gap between what present tools provide and what the technology will deliver.

It is then clear that we need improvements of the design quality and designer's productivity. This may be achieved in two ways that can be combined:

- Providing higher level design tools allowing to start from a higher level of abstraction. After the success and the widespread acceptance of logic and RTL synthesis, the next step is behavioral synthesis, commonly called architectural or high-level synthesis.

- Using more structured design methodologies allowing for an extensive reuse of existing components and subsystems. It seems that 70% of new designs correspond to existing components that cannot be reused because of a lack of methodologies and tools.

1.2 FROM PHYSICAL DESIGN TO SYSTEM DESIGN: ABSTRACTION LEVELS

The arrival and acceptance of standard Hardware Design Languages (HDLs) such as VHDL and VERILOG, have promoted high level specification of electronic circuits. HDLs may be used for the specification of whole systems, as well as subsystems that may then be assembled within a hierarchical (structural) description. The fact that various tools (for synthesis and simulation) have been developed this last decade has also helped high level specification for VLSI to emerge and to gain more and more acceptance in the VLSI design community [MC80].

HDLs can be used for design specification at various abstraction levels, from gates to the behavioral level. Timing concepts will be used to fix the abstraction level of design specifications.

Timing is the main issue during the process of designing an integrated circuit. In fact regardless of the abstraction level, the design process may be defined as the refinement of high level concepts (operations, primitives, statements or constructs) into lower level concepts using more detailed timing units. Table 1.1 illustrates this concept.

At the lowest level the basic timing unit is the delay. The design is specified in terms of gates and devices that are interconnected through nets. A typical specification at this level is a schematic. Of course this may be described using a HDL format such as VHDL or VERILOG. At this level the components of the design (gates, nets, devices) are characterized by delays. Simulation and timing analysis tools are needed in order to compute the performance of the full design. The clock period will correspond to the longest path between two memorization elements. A path may include several components.

Introduction

Description Level	Time Unit	Primitives	Description Organization
Physical Level	Delay	Gates, Devices	Schematic
Logic and RT-Level	Clock Cycle	Registers, Operators, Transfers	Boolean equations, BDDs, FSMs
Algorithmic Level	Computation step	Computation, Control	BFSMs, DFG, CFG
System-Level	Transaction	Processes, Communication	Communicating Processes

Table 1.1 Timing Concepts and Specification Levels through the Design Process

The next level is called logic or Register Transfer Level (RTL). The design is specified at the clock cycle level. Typical descriptions will state what to do at each clock cycle. A description is generally formed of a set of registers, operators and transfers between registers and operators. A typical representations used by synthesis tools at this level are boolean equations, FSMs and BDDs. Such representation can be extracted automatically from HDLs. The main design steps applied starting from this level are logic optimizations, synthesis state encoding and technology mapping. The main role of these transformations is to fix the clock period and the gate count. The logic optimization is generally a trade-off between these two parameters. In VHDL, when writing an RTL description we assume that all the computations between two wait statements can be transformed into a set of gates able to perform that computations within a clock period. The decomposition of the clock period into delays is made automatically by the RTL synthesis. Of course the designer can control the synthesis results through different writing styles or a set of constraints. This kind of description is also called *synchronous* description or *cycle based* description.

The next level is called behavioral or algorithmic level. A design is specified in terms of computation steps. The concept of operations and control statements (loop, wait) will be used to sequence the I/O events and the computation. At this level we have an event driven specification. A typical description will be composed of a set of protocols to exchange data with the external world and an internal computation. A computation step is composed of the set of operations executed between two successive I/O and/or synchronization points. A computation step may take several clock cycles. The main function of behavioral synthesis is to split these computation steps into a set of clock cycles. Moreover, some of these computation steps may include data dependent computation implying a non predictable (variable) computation time. A typical representation at this level is Behavioral Finite State Machine (BFSM), Control Flow Graph (CFG), Data Flow Graph (DFG) and Control-Data Flow Graph (CDFG). When writing a behavioral description we assume that all the computations between two synchronization points can be decomposed into a set of clock cycles that respects the communication protocol. The decomposition of a computation step into clock cycles is made automatically by behavioral synthesis. One of the major problems when using behavioral synthesis is to specify complex and precise protocols within behavioral description.

At the highest level, we have the system level specification. Such a specification includes distributed control and multi-thread computations. The basic timing unit at this level is the communication transaction. The basic primitive is the process. A description will be composed of a set of hierarchical, parallel and communicating processes.

Most hardware description languages allow the three first abstraction levels (Gates, RTL, Behavior). For each of these languages each level corresponds to a writing style using a subset of the language. One can note that we can describe and simulate a full system at the gate level or describe a gate at the behavioral level.

When describing large systems, it is often the case that all the specification levels have to be combined. In fact a system specification is seldom given in a unique specification level. Generally it is composed of blocks described at different abstraction levels. For example, figure 1.1 shows the block diagram of a system describing an answering machine. The answering machine acts between a line and a phone. It is connected to these environments through a D/A, A/D converter. It is also connected to two tape controllers. The tapes are aimed to contain the announce of the answering machine owner (e.g. Thanks for calling, I can't answer the phone right now, Please leave a message after the tone) and the messages left by the callers. A Timer is used to provide a time reference. The main block is a central controller aimed to synchronize the communication between the different components and the operators.

It is easy to see that in order to describe such an example we may need several abstraction levels. For example, the A/D and D/A components will be described at the physical level, the tape controllers and the timer will be described at the clock cycle level (RTL) and the controller may be described at the behavioral level. Since we have several description levels we may need

Introduction

different design tools acting at different levels. Physical level synthesis tools (schematic capture, place and route, ...) will be used for the design of the A/D parts. RTL tools (e.g. logic synthesis) will be used for the design of the other interface components. Higher level design tools will be used for the design of the controller.

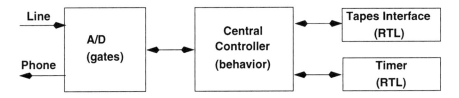

Figure 1.1 A Multi-level Description System (Answering Machine)

1.3 BEHAVIORAL SYNTHESIS

Behavioral synthesis is the process that starts from a behavioral or functional specification (e.g. an algorithm) and produces an architecture able to execute the initial specification. The architecture is generally given as a Register-Transfer Level (RTL) specification and is composed of a datapath and a controller. The behavioral description specifies the function to be performed by the design. It may be textual or graphical. A behavioral synthesis tool acts as a compiler that maps a high level specification into an architecture. In order to modify the architecture you simply have to change the behavioral description and return it through the behavioral synthesis tool. The use of behavioral synthesis induce a drastic increase in productivity since behavioral descriptions are smaller and easier to write and modify.

High level synthesis is the bridge between what is called system design and CAD tools acting at the logic and RT Levels.

As shown in figure 1.2 a full system design automation methodology will start from a system level specification. This may be given in a language such as SDL, Esterel or StateChart. The first step is in charge of the system level design and synthesis. It is aimed at partitioning and communication synthesis.

The result of the partitioning is an heterogeneous architecture composed of a set of interconnected subsystems. Each part can be defined using specific description language and designed using appropriate tools and methodology.

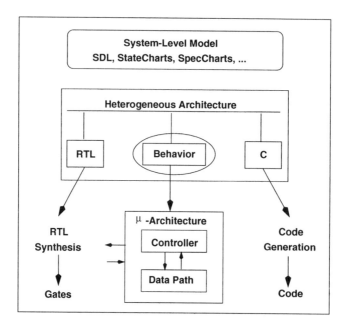

Figure 1.2 System Design Process

After partitioning, algorithmic synthesis will handle subsystems described at the computation step level or behavioral level. This step will decompose each behavioral subsystem into a controller and a datapath. More generally, the results of behavioral synthesis is a micro-architecture composed of controllers, datapaths and glue cells. Here only, RTL and logic synthesis tools can be used.

As we can see, behavioral synthesis is an important tile in system design. However, this should be combined with other synthesis tools such as code generation and RTL synthesis in order to design full systems. We will also see later that even for behavioral synthesis we may need different tools to cope with different application domains.

1.3.1 Behavioral Synthesis Tools

A plethora of behavioral synthesis tools have been published in the literature [CPTR89, LG88, PKG86, TDW+88, Pan88, KNRR88, Zim79, Pen86, AP89, GKP85, MK88, LMWV91]. Only few of them have been applied for the design of VLSI chips [MCG+90, JPO93, Inc94, TDW+88, LHCF96]. While behavioral synthesis tools have been applied successfully to DSP algorithms, arithmetic computation and interfaces, behavioral synthesis was less successful for the design of real-time controllers, complex heterogeneous design and data dependent computation. None of the existing behavioral synthesis tool has been recognized as a universal tool that may be efficient in all application domains and for all kind of architectures. Existing behavioral synthesis tools may differ from several points of views. These correspond to the main choices that have to be made when designing a behavioral synthesis tool and more generally application generators [Cle88]:

1. Application domain: It fixes for which kind of application the tool can be used e.g. DSP operators, controller, interfaces, ...

2. Underlying design model: This is generally defined by the intermediate form used for the refinement of the behavioral specification into architecture. This may be language oriented design models (data and/or control flow graphs) or architecture oriented models (FSM based). These models are strongly related to application domains.

3. Flexibility of the synthesis process: This fixes the parts that are under the designer's control (e.g. synthesis options) and the parts that cannot be changed (e.g. optimization algorithm).

4. Input description: This may be textual or graphical. It may include dialogues with the user.

5. Output or architecture: this is generally made of a controller and a datapath. The architectural style should be suitable for the application domain.

6. The synthesis algorithm and the synthesis flow: Most behavioral synthesis tools include the same steps: operation scheduling, resources allocation, resource binding and control generation. However the algorithms used for each of these steps may be different from one tool to another. For instance scheduling may be implemented following an ASAP, ALAP or list scheduling. Besides the order of these steps may differ for each tool. Scheduling and allocation may be applied in different order.

All the above mentioned criteria are interrelated. This makes it very difficult to compare existing behavioral synthesis tools.

1.3.2 Behavioral Synthesis and Application Domains

The restriction of a behavioral synthesis tool to a specific application domain allows to reduce the complexity of the synthesis process and to produce more efficient results. For behavioral synthesis two application domains are generally distinguished in the literature, data flow oriented applications and control flow oriented applications. A data flow oriented application acts on regular data streams. The inputs and outputs of the design are made of regular data streams. The behavior is generally specified as a periodic set of operations that have to be executed on each new piece of data. Inputs and outputs are signals with fixed throughput (figure 1.3). A filter is a typical data flow oriented application.

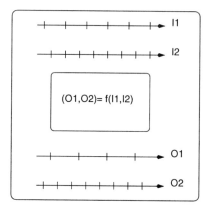

Figure 1.3 A typical data flow oriented application

Control flow applications are driven by a set of commands that have to be interpreted. Each command (sequence of control/data) may imply reading or writing complex data structures and the execution of a specific algorithm related to the command. The computation performed may be data dependent. This includes handshaking with the environment, data dependent loops and fast reaction to interrupts. For the design of complex systems, it is often the case where data flow oriented components have to be mixed with control flow oriented components.

A typical mixed control-data-flow design is shown in figure 1.4. It is composed of a controller and an execution part made of two memories (M1, M2) and a DSP coprocessor. The latter executes regular computation (e.g. DCT, IDCT, FILTER, ...) on data streams provided according to a fixed throughput by

Introduction

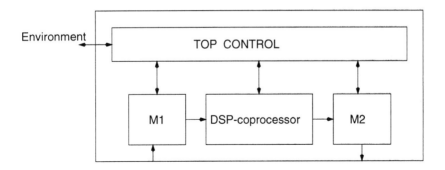

Figure 1.4 A typical DSP application: mixing data flow oriented and control flow oriented designs

M1. It also outputs a regular data stream to M2. The top controller is in charge of synchronizing the three components according to the information provided by the environment. The function of the top control may include sophisticated operations such as reading/writing a local memory from/to an external memory accessed through a specific protocol. This may be done in parallel with the coprocessor execution.

In order to handle complex design mixing both control and data flow oriented computations, a three steps modular approach is needed:

1. Partition the design into control flow oriented modules and data flow oriented modules.
2. Use a data flow oriented synthesis tool for the design of the data flow modules.
3. Use a control flow oriented synthesis tool for the design of the top control. During this step the data flow modules will be used as a component.

The use of behavioral synthesis within a modular design approach is explained in chapter 4. An example of modular design is given in chapter 7.

1.3.3 Design representation for behavioral synthesis

Behavioral synthesis is the task of refining a behavioral description into a Register Transfer Level (RTL) description through a set of successive transformations. The two main models used during this refinement process are the internal representation of the behavioral description and the target architecture. The internal representation, also called intermediate form, fixes the underlying design model of the behavioral synthesis tool. The choice of this model constitutes one of the main decisions when building a behavioral synthesis tool. Although intermediate forms are generally not accessible to the users of such tools, their understanding may help to better comprehend the functioning of the tool.

There are mainly two types of intermediate forms, the first are language oriented and the latter are architecture oriented. Language oriented intermediate forms use a graph representation. The main employed representations are Data-Flow Graph (DFG), Control-Flow Graph (CFG) and Control-Data-Flow Graph (CDFG) which mixes both previous representation styles. The architecture oriented intermediate forms use an FSM/Datapath representation which is closer to behavioral synthesis output.

The use of language oriented intermediate forms makes easier the application of high-level behavioral transformations. However, the use of architecture oriented models allows to handle more sophisticated architectures with features such as control pipelining and component reuse. These intermediate representations will be detailed in chapter 2.

1.3.4 Flexibility of the Synthesis Process

Like all generation/transformation applications a behavioral synthesis tool may be characterized by invariant (or fixed) parts and variant (extendible or exchangeable) parts. The latters fix the flexibility of the synthesis process. An invariant part is a feature that cannot be changed e.g. fixed assumption about the implementation. These should be restricted to design details that users prefer not to worry about [Cle88]. Organizing the target architecture as a datapath/controller model is an invariant part for most behavioral synthesis tools. A variant part is a feature under the user's control. Variant parts correspond to parameters and extensions to the behavioral synthesis process. The most common form of parameterization is the use of a library of abstract components that can be mapped onto several technologies. A component implementation

Introduction 11

is a variant that the designer fixes when the component is instantiated. The component has fixed parts, e.g. the interface that cannot be changed. The flexibility of a behavioral synthesis tool may also be increased drastically by allowing the instantiation of user defined components. For instance, the co-processor in the FSMC model (see chapter 2) constitutes an extension of the behavioral synthesis tools:

- The input description language is extended with new operations (procedure and function calls).

- The architecture is extended with co-processors.

Adding flexibility to behavioral synthesis may have some drawbacks such as decreasing testability, analysis capability and predictability of the results. This comes from the fact that behavioral synthesis cannot efficiently analyze the user defined extensions. However, despite these drawbacks, flexibility allows to dramatically increase the applicability of a behavioral synthesis tool.

1.3.5 Behavioral Description

Once the application domain defined, the design representation selected and the flexibility issues fixed, a behavioral model description language may be selected. Of course, this language has to be compatible with the previous choices. Regardless the language selected (textual/graphical, data flow/procedural, ...) most behavioral synthesis tools mix a description language and a command language. The description language may be textual (e.g. VHDL [VHD87], Verilog [TM91], SILAGE [Hil85]) or graphical (e.g. StateCharts [H+90]). Some of these languages (e.g. SDL [SDL87]) provide both textual and graphical entries. The command language may be a textual format describing scripts for synthesis or dialogue in an interactive approach. Although very important from the designer's point of view, the behavioral description language is of minor importance for the behavioral synthesis tool builder. In fact, the first step of most existing behavioral synthesis tools is to map the input description onto an intermediate form. However, a behavioral description language should be selected carefully in order to fit well with the application domain and the internal representation used by the behavioral compiler.

1.3.6 Architectural model

Behavioral synthesis tools produce an architecture. This is generally coded as an RTL description (VHDL, VERILOG) that can be processed by RTL and logic processing tools. Most behavioral synthesis tools produce an architecture made of a datapath and a controller. However within this model, different styles may be used. The main options:

- Datapath style: There are two main datapath styles depending on the way the communication between the components of the datapath is organized. The communication between registers, operators and external parts may be made through buses or through MUXs.

- Controller style: The two main styles used by behavioral synthesis for organizing the controller are the flat FSM model and the microcoded model.

- Pipelining:There are three kinds of pipelining: control pipelining, datapath pipelining, and component pipelining [GR94]. Pipelining decomposes a computation into stages in order to allow the different stages to be executed concurrently. In control pipelining, the datapath and the controller are organized to work concurrently on subsequent instructions. In datapath pipelining, a part of the datapath is duplicated in order to allow parallel execution of subsequent statements. In component pipelining, the computation of a given component is pipelined in order to increase its utilization.

- Component model: High level synthesis abstracts the technology as a set of components (registers, MUXs, operators). The model of these components may be fixed or flexible. In the latter case, a component modeling language is used. In this case, the component model may be changed without changing the behavioral synthesis tool itself.

All these architectural models and styles will be detailed in chapter 2.

1.3.7 Algorithms for behavioral synthesis

Behavioral synthesis produces an architecture starting from a functional specification. It is generally decomposed into five major transformations: generation of an intermediate form, scheduling, allocation, binding and architecture generation. Only a brief outline of these steps will be given in this section. An

Introduction

excellent overview of these techniques may be found in [Mic94]. Figure 1.5 illustrates the synthesis steps through a simple example. Each design step may produce several solutions. Figure 1.5 shows two alternative solutions for each synthesis step. At each step the solution framed with bold lines will be selected to be used as input for the next step.

1. **Compilation of the behavioral description and generation of an intermediate form**

   This step may include compiler-like transformations (called high-level transformations in the literature [WT87]) aimed at removing all the details related to the description language and the writing style. Such transformations include constants propagation, dead code elimination, loop unfolding and procedure expansion. When a language oriented intermediate form is used, this step is straight-forward since the graph models are very closer to most behavioral description languages. But when an architecture oriented intermediate form is used, this step makes use of complex transformations in order to transform the behavioral description into an FSM. These may include partial scheduling and allocation. In the example of figure 1.5, a data flow graph representation will be used during the synthesis process.

2. **Scheduling**

   Scheduling is the partitioning of the behavioral description into subgraphs, each one being executed in a single control step. A control step corresponds to a transition of an FSM. It may include several operations to be executed in parallel. Of course, this assumes that enough resources will be allocated in order to perform the parallel computations. Some scheduling algorithms are performed under resource constraints and/or timing constraints. Different scheduling models will be detailed in chapter 3. In the example of figure 1.5, two schedules are produced. In the first solution (left), two control steps (CT1,CT2) are required for the execution of the behavioral description. However, in this solution, the execution of CT1 would require two operators in order to execute the two parallel operations (+,-) performed in CT1. The second solution (right) requires three control steps to execute. However, no parallel operations are performed.

3. **Allocation**

   Allocation fixes the amount and types of resources needed to execute the behavioral description. This step fixes the number and type of the compu-

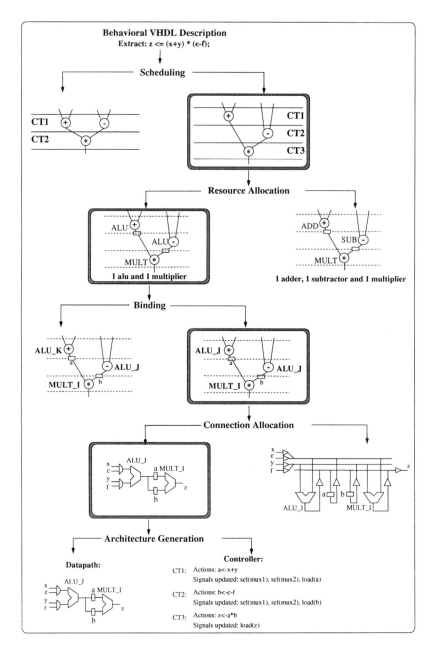

Figure 1.5 Behavioral synthesis steps

tation units operators (e.g. ALU and co-processors in the case of FSMC), the storage units (registers, register files) and connection units (MUXs, buses, wires). Of course, the amount of resources may restrict the concurrency allowed in the datapath and restrict the scheduling. In the example of figure 1.5, starting from the scheduled description represented as a DFG, the resource allocation task needs to allocate operators to execute the operations of the behavioral model and registers to store the intermediate values which are used in more than one clock period. In this case also, two solutions are produced. In the first one, two types of operators are allocated (ALU, MULT). In the second solution, three kinds of operators are allocated (ADD, SUB, MULT).

4. Binding

This step decides which resources will be used by each operation of the behavioral description. In several systems, this step is included within the allocation step. In the example of figure 1.5, two binding solutions are shown. In the first one, the operations (+,-) are assigned to two different ALUs. While, in the second, the two operations are executed on the same ALU. The output of the two first operations are stored into two registers (a,b).

5. Connection Allocation

This step fixes the resources needed for the communication between the units of the datapath (registers and operators). According to the architecture style, this step will generate the paths needed for data exchange. For the example of figure 1.5, 2 solutions are shown. In the first, the communication paths are made using multiplexers. In the second, the communication paths use buses. Access to buses is regulated by the use of switches (tri-state components). The resulting datapath includes all the needed paths for data transfers between inputs and outputs, internal registers and operators.

6. Architecture generation

This step produces an RTL description of the synthesis design. The controller and the datapath are built starting from the result of the previous steps. In the example of figure 1.5, starting from the multiplexer based datapath, a datapath controller model is generated. The controller includes the sequencing of the control steps and the activation of the datapath components during each control step.

The synthesis process detailed in figure 1.5 is only an indicative model. Most existing behavioral synthesis tools make use of different models. The main difference between the existing synthesis schemes is the order of scheduling and allocation. Although scheduling and allocation are 2 distinct tasks, their performances are closely related [MPC90]: whenever scheduling is performed before allocation, additional constraints are imposed on the scheduled operations with respect to allocation; on the other hand if allocation is performed before scheduling, restrictions are imposed on the scheduling step. As a result, much controversy has been expressed about their ordering. This inter-dependency between scheduling and allocation became harder to handle when generic and complex operations have to be managed. When multi-cycling is allowed within the execution scheme of the functional unit operations, the scheduling gives the latency for the execution of the different operations. Once scheduling is done, the constraints upon the allocation task are more severe and proceeding with the synthesis shall not offer as many solutions as it may be expected with the initial unscheduled description.

In order to deal with the inter-dependency between the two synthesis subtasks, different solutions have been implemented in several synthesis tools. Many of them perform scheduling before allocation, among other tools: CALLAS [BK+92], EASY [Sto91], HAL [PKG86], HIS [CBH+91]. These synthesis systems usually execute scheduling under constraints, such as the maximum number of functional units allowed. Other tools perform allocation before scheduling according to their design-flow; a few examples of such tools are: CATHEDRAL [N+91], GAUT [MSDP93], VSS [LG88]. In the latter case, once allocation has been achieved (as well as the functional unit binding), the scheduling is constrained and much parallelism may be lost. Consequently, the advantage of scheduling a description before undertaking allocation is to have a simplified allocation task, because in this case allocation became a hardware sharing operation once scheduling has been performed under the assumption of unlimited hardware. Unfortunately this also means limited possibilities and thus limited optimizations for the allocation. Moreover within such a design flow, dealing with multi-cycle operations makes scheduling a complex and tedious task. On the other hand, allocating the resources before scheduling simplifies this last operation. Since scheduling is not known when allocation is done, this solution tends to produce maximum hardware. In fact the problem in this case results from the difficulty encountered to decide upon the hardware sharing before allocation. Therefore an allocation step is required to enable hardware sharing.

An ideal design-flow would be to perform scheduling and allocation simultaneously. The formulation of this problem has been done through the integer linear programming (ILP) [GDWL92], which corresponds to an optimization problem

Introduction

of resource constrained scheduling. This problem has one major inconvenience: the decision problem is NP complete and can therefore not be used for large examples. The behavioral synthesis tool AMICAL (see chapter 5) makes use of a synthesis flow when both scheduling and allocation are decomposed into two basic steps that performs partial scheduling or allocation. The synthesis process proceeds in alternating scheduling and allocation steps. It starts with a partial scheduling that reorganizes the input description into a set of control steps (called macro-cycles). This step produces an FSMC. A partial allocation (including functional unit binding) is then performed. It selects for each action a functional unit that will undertake it. Another scheduling step is then achieved at the clock cycle level, before final allocation step, needed in order to allocate the communication.

1.4 DESIGN REUSE

Design reuse is the use of existing design knowledge or artifacts to build new design artifacts [FF95]. This includes component reuse and design process reuse.

A component may be a part of the system to be designed or a design element needed by the design process (e.g. test vectors or a testbench). Component reuse includes:

- Building reusable components.
- Building new designs with reusable components.

A component may be used in different systems e.g. a functional module used in two different designs. A component may also be reused for the design of the same system:

- several instances of a given subsystem.
- during architecture design and system planning, a component may be reused in several alternative solutions.
- a component may be used at different steps of the design process, e.g. using the same testbench for the simulation of the design at several abstraction levels (RTL and gate-level or RTL and behavioral level).

Reuse can also be extended to the reuse of the process deployed to produce a design. This can range from a simple parameterized generator (e.g. generators for RAM, ROM, ALU, multiplier ...) to an application specific synthesis tool. To some extent the use of a behavioral compiler is a kind of design reuse [MMM95].

Component reuse is a promising technology that may bring a drastic increase in productivity and quality. Then one may ask why is reuse still not a very common practice?

A recent study [FF95] on software reuse has shown several truths that contradict some common wisdoms:

- Designer would prefer to reuse rather than building for scratch.
- The language used is not determinant for promoting the reuse. Then the existence of an OO-HDL may not have a large impact on reuse in hardware design.
- Reuse does not increase with designers' experience.

Lack of education is the main factor that restricts design reuse. It seems that the main factors that may promote reuse are economic and education. The situation for hardware is more promising than on the software side. In fact the economic reasons are evident while the education is quite large for the design at the physical and logic levels. However, we are still lacking techniques and methods for hardware component reuse at the behavioral level.

1.5 COMPONENT REUSE IN VLSI

Component reuse is generally associated to structured or modular design methodology. This methodology has two main concepts: hierarchy (or modularity) and regularity [TRLG81]. Hierarchies are used for decomposing a complex design into sub-parts that are more manageable. Regularity is aimed to maximize the reuse of already designed components and subsystems.

This kind of methodology for VLSI was introduced first by Mead and Conway [MC80]. It has proven its efficiency in several other domains of system design [Seq83]. Each time a new design level is adopted, the design community starts looking for structured design methodologies acting at that level in order to master complexity. Structured design methodology has been adopted since the starting of the domain of VLSI design [MC80]. This methodology has been applied to the physical level with structured design tools [TRLG81, Seq86]. It has also been applied to logic and register transfer level design [GC93].

At each level, different techniques are used for hierarchical decomposition and for component specification and reuse. At the register transfer level a typical environment will provide methods for modeling generic operators (such as adder, multiplier, ALU ...) able to execute basic operations (for example: +, *, -, ...). These operators, acting as black boxes in the netlist (output description), are a hybrid between arithmetic operator and library cell. They are the link between the HDL operators and the components of the final library. Each operator corresponds to a unit that may perform one or several functions. Since the global description is given at the clock cycle level (the basic time unit), the execution timing delay of each of these operations cannot exceed a clock cycle.

At the behavioral level we may have operations that require several clock cycles for execution and others may invoke handshaking for parameter passing. Then, the interface of functional units will be a set of protocols allowing to communicate with the behavioral description. For example a memory may be modeled as a unit that can execute read and write operations. The unit will be accessed only through these two operations. The internal organization of the memory (including the addressing functions) is hidden.

In order to apply modular design methodologies at the behavioral level [GDWL92], we need to scale up the concept of the operator in order to allow the execution of complex operations or modules, such as multi-cycles operations and operations with complex parameter passing protocols. This implies that we will need more complex functional units allowing for:

- Internal memory (for example: local accumulators).
- Different timing schemes for different operations (for example: in the case of a memory unit where a read operation takes 1 cycles and a write operation takes 2 cycles).

1.6 MODULAR DESIGN METHODOLOGY FOR COMPONENT REUSE AT THE BEHAVIORAL LEVEL

Modular design methodology allows to handle very complex design with structured approach. Modular design proceeds by partitioning a system into modules. The implementation details of these modules are hidden. Proper partitioning allows independence between the design of the different parts. The decomposition is generally guided by structuring rules aimed to hide local design decisions, such that only the interface of each module is visible.

Structured design methodology is based on the divide-and-conquer principle. Regardless the abstraction level, the structured design methodology for VLSI consists of three main steps in the design-flow, as illustrated by figure 1.6:

1. Partitioning can be applied on the system specification in order to split the system into simpler subsystems or modules.

2. Each module thus generated can be designed independently using a specific library of components.

3. The abstraction of the subsystem has to be done in order to enable its reuse as a complex library element.

For the application of structured design methodology to behavioral synthesis, the reuse of subsystems that have been synthesized during a previous high-level synthesis session has to be allowed.

Partitioning and subsystem design are closely inter-related. The hierarchical decomposition may be influenced by the set of already existing components. On the other side, the selection of the components is influenced by the hierarchical decomposition of the design. The power and flexibility of such scheme depend on the range of components that may be used in the library.

When such a methodology has to be automated, the main issue is to find the right strategy for component abstraction: description for reuse. The library may contain (dashed arrows in figure 1.6):

- Components and subsystems that may be designed using other design methods and tools, and

Introduction

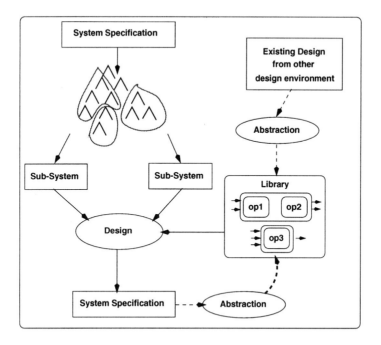

Figure 1.6 Use of Modular Design Methodology

- Subsystems that result from an early synthesis process.

The choice of the components is very important during the design process. The use of simple components (e.g. standard functional units) may increase the size of the behavioral description as well as that of the resulting controller. On the other side the use of more complex components may increase the size of the resulting datapath.

Chapter 5 presents AMICAL a behavioral synthesis tool for design reuse. The main features that make AMICAL suitable for design reuse are:

- A powerful internal design representation (FSMC) for complex operations.
- A flexible synthesis model allowing user defined components in addition to standard components.

1.7 SUMMARY

This chapter discussed the need for combining behavioral synthesis and design reuse in order to improve designers' productivity and design quality.

Behavioral synthesis is emerging after the success and widespread of logic and RTL synthesis. In order to show the main differences between abstraction levels, section 1.2 discussed the different concepts used when modeling systems at the gate, RT, behavioral and system levels.

Behavioral synthesis is the process that starts from a behavioral or functional specification and produces an architecture able to execute the initial specification. Existing behavioral synthesis tools may differ from six points of view: the application domain, the underlying design model, the flexibility of the synthesis process, the input description, the target architecture and the synthesis algorithms.

Design reuse is the use of existing components to build large systems. It is generally associated to modular design. Design reuse for VLSI was introduced first by Mead and Conway [MC80] for the physical level. Each time a new design level is adopted, the design community starts looking at design reuse strategies at that level.

At the behavioral level, components stand for large subsystems. The main steps involved in a design methodology for reuse include system level analysis and partitioning, subsystem design using behavioral synthesis and subsystem abstraction for reuse.

2
MODELS FOR BEHAVIORAL SYNTHESIS

Behavioral synthesis is the task of refining a behavioral description into a Register Transfer Level (RTL) description through a set of successive transformations. The two main models used during this refinement process are the internal representation of the behavioral description, and the target architecture. The internal representation, also called intermediate form, fixes the underlying design model of the behavioral synthesis tool (see CFG, DFG or CDFG further). The choice of this model constitutes one of the main decisions when building a behavioral synthesis tool. Although intermediate forms are generally not accessible to the users of such tools, their understanding may help to better comprehend the functioning of the tool. The target architecture is generally composed of a datapath and a controller; when fixed, it also fixes the performances of the synthesized design. The understanding of the target architecture is particularly important in order to evaluate and debug the result of behavioral synthesis.

This chapter gives an overview of both models used in behavioral synthesis.

2.1 DESIGN REPRESENTATION FOR BEHAVIORAL SYNTHESIS

Behavioral synthesis is a transformation process generally based on a well defined internal model. This is generally called intermediate form. The different steps of the behavioral synthesis, from input description to architecture, can be explained as a transformation of this internal model. There are mainly two kinds of intermediate forms : the first are language oriented and the latter are architecture oriented. Both may be used in order to achieve the behavioral synthesis. Language oriented intermediate forms use a graph representation. The main employed representations are Data-Flow Graph (DFG), Control-Flow Graph (CFG), and Control-Data-Flow Graph (CDFG) which mixes both previous representation styles. The architecture oriented intermediate forms use an FSM/Data path representation which is more closed to the output of behavioral synthesis.

2.1.1 Language oriented intermediate forms

Various representations have appeared in the literature [KD92, CT89, Wol91, Sto91, Jon93, GW92], mostly based on flow-graphs. The main kinds are Data, Control and Control-Data flow representations as introduced above.

Data-Flow Graph

Data flow graphs are the most popular representation of a program in high level synthesis. Nodes represent the operators of the program, edges represent values. The function of node is to generate a new value on its outputs depending on its inputs.

Definition 2.1 (Data-Flow Graph) *A data-flow graph DFG is a graph $G = (V, E)$, where:*

> *(i)$V = \{v_1, \ldots, v_n\}$ is a finite set whose elements are nodes, and*
> *(ii) $E \subset V \times V$ is an asymmetric data flow relation, whose elements are directed data edges.*

The nodes in a DFG represent operations. A directed data edge eij from $v_i \in V$ to $v_j \in V$ exits if the data produced by operation o_i (represented by v_i) is consumed by operation o_j (represented by v_j). v_i is said to be an immediate predecessor of v_j. Similarly, v_j is said to be an immediate successor of v_i.

A data flow graph example is given in figure 2.1, representing the computation $e := (a+c)*(b-d)$. This graph is composed of three nodes v_1 representing the operation $+$, v_2 representing $-$ and v_3 representing $*$. Both data produced by v_1 and v_2 are consumed by v_3.

Models for behavioral synthesis

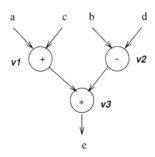

Figure 2.1 Example of a simple data flow graph

In the synchronous data flow model we assume that an edge may hold at most one value. Then, we assume that all operators consume their inputs before new values are produced on their inputs edges.

In the asynchronous data flow model, we assume that each edge may hold an infinite set of values stored in an input queue. In this model we assume that inputs arrivals and the computations are performed at different and independent throughputs. Only the synchronous data flow model is used in behavioral synthesis. This model is powerful for expression representation. However it is restricted for the representation of control structures.

Control-Flow Graph

Control flow graphs are the most suited representation to model control design, which contain many(possibly nested) loops, global exceptions, synchronization and procedure calls; in other words, features that reflect the inherent properties of controllers.

Definition 2.2 (Control flow Graph) *A control flow graph CFG is a graph* $G = (V, E)$, *where :*

> *(i)* $V = \{v_1, \ldots, v_n\}$ *is a finite set whose elements are nodes, and*
> *(ii)* $E \subset V \times V$ *is a control flow relation, whose elements are directed sequence edges.*

Figure 2.2(a) show an algorithmic description of a GCD (Greatest Common Divisor). The corresponding CFG is given in figure 2.2(b). It consists of seven

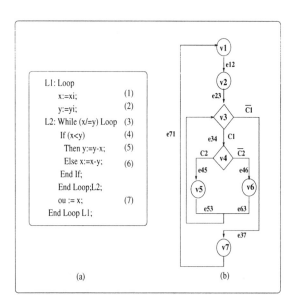

Figure 2.2 (a) Algorithmic description of GCD. (b) CFG of the GCD example

nodes, $V = \{v_1, \ldots, v_7\}$ and nine edges: $E = \{e_{12}, e_{23}, e_{34}, e_{45}, e_{46}, e_{53}, e_{63}, e_{37}, e_{71}\}$ Four edges are weighted. We can easily see that all these edges are issued from branch nodes. We assume that this description corresponds to a circuit running indefinitely, that is the reason e_{71} is inserted.

Definition 2.3 (Graph node) *Nodes are of two classes:*

- *operation nodes such as assignments, logic, arithmetic operations and procedure calls, represented by the subset of V, Vo. For the example of figure 2.2,*
  $Vo = \{v_1, v_2, v_5, v_6, v_7\}$.

- *branch nodes modeling conditional statements such as If, Case and Conditional loops, represented by Vb. For the example of figure 2.2, $Vb = \{v_3, v_4\}$.*

Thus $V = Vo \cup Vb$

The operations nodes have only one successor, while branch nodes may have more than one successor.

Definition 2.4 (Graph edge) *An edge represents the precedence relation between two nodes v_i and v_j. An edge e_{ij} is weighted by a condition $Cond_{ij}$, meaning that the operation represented by v_j will be executed if v_i is executed and $Cond_{ij}$ is evaluated to True.*

For this reason the nodes have to satisfy the following two properties:

Property 2.1 *if v_i is an operation node and v_j is an immediate successor of v_i, i.e $(v_i, v_j) \in E$ then: $Cond_{ij} = 1$.*

Property 2.2 *if v_i is a branch node, and $(v_{i1}, \ldots, v_{ik})$ are the k immediate successors of v_i then:*

$Cond_{i,im} \wedge Cond_{i,ih} = 0$, *for all* $m, h \in \{1, \ldots, k\}$ *and* $m \neq h$, *and* $\sum_{l=1}^{k} Cond_{i,il} = 1$.

This property insures that the CFG is deterministic.

While this kind of graph models well control structure including generalized nested loops combined with synchronization statement (wait), control statements (if, case), and exceptions (EXIT), it provides restricted facilities for data flow analysis and transformations.

Control-Data-Flow Graph

The most common representation used for behavioral synthesis is the CDFG [PK89]. This model extends DFG with control nodes (If, case, loops). This model is very suited for the representation of data flow oriented applications. It is used in BC [Inc94], CATHEDRAL [MCG+90], ... It is also used by several control flow oriented synthesis tools such as HIS [CBH+91], HERCULES [MK88], CALLAS [BK+92] and BC [Inc94].

We distinguish two main representations, the first extending data flow graphs to support control constructs by adding new types of nodes such as branch and merge nodes. As defined in [Sto91, Jon93], a branch node is a conditional node. The behavior of such a node is to pass data from the data input port to one of the output ports. The output port is selected according to the value of control ports. Similarly, the behavior of a merge node is to pass data from one data input port to the output port, the input port is selected according to the value on control ports. So, control structure is embedded within the data flow graph. Figure 2.3 shows the extended control-data flow graph for the GCD example, x'i and y'i represent the computation of x and y for each iteration. .

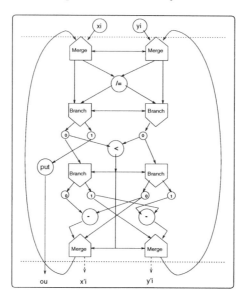

Figure 2.3 Combined control-data flow graph for GCD example.

The second type of control data flow graph representation deals with basic blocks which represent sequences of operations in which no control flow is present. For each basic block, a data flow graph is made. In this scheme, the control flow nodes are kept a part from the data flow nodes, so a basic block represents a control statement node and a data flow graph with several data flow nodes. The CDFG using basic blocks for the GCD example is shown in figure 2.4. This representation is used in [OG86] and [LG88]. One variation of this model is the polar hierarchical acyclic graph used in the Hercules high-level synthesis system [MK88]. The nodes in this graph represent operations and the edges represent the dependencies between the operations. The hierarchy supports procedure call, conditional branching and loops. This means that the bodies of such constructs are in separate graphs.

Models for behavioral synthesis 29

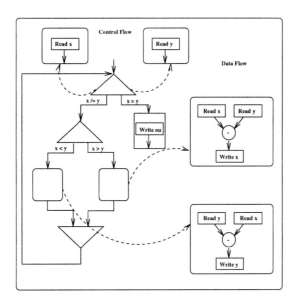

Figure 2.4 Separate control-data flow graph for GCD example.

2.1.2 Architecture Oriented Intermediate Forms

This kind of intermediate form is more closed to the architecture produced by behavioral synthesis than to behavioral description. The data path and the controller may be represented explicitly when using this model. The controller is generally abstracted as an FSM. The data path is modeled as a set of assignment and expressions including operation on data. The main kind of architecture oriented forms are FSM with data path model(FSMD) defined by Gajski [GDWL92] and the FSM with co-processors(FSMC) used by AMICAL.

The FSMD representation

The FSMD model was introduced by Gajski [GDWL92] as a universal model that represents all hardware design. An FSMD is an FSM extended with operations on data. It is defined as a set of variables V, a set of expressions $E = \{f(x,y,z,\ldots)|x,y,z \in V\}$ and a set of storage assignment $A = \{x <= e | x \in V, e \in E\}$. Besides, status SS is defined in order to store the execution report of expressions.

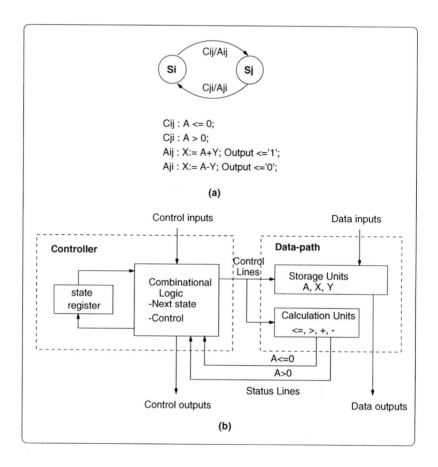

Figure 2.5 (a)FSMD Model, (b) FSMD implementation

Definition 2.5 (FSMD) *An FSMD is formulated as a quintuple:*

$< S, I \times SS, O \times A, f, h >$, *Where:*

- *S : the set of states of the FSMD*
- $I \times SS$: *The set of inputs of the FSMD. Inputs extended with status expressions*
- $O \times A$: *The set of outputs of the FSMD. Outputs extended with variable assignments.*

Models for behavioral synthesis 31

- f : *The next state function, mapping* $S \times (I \times SS) \to S$
- h : *The output function, mapping* $S \times (I \times SS) \to (O \times A)$.

The FSMD computes new values for variables stored in the data path and produces outputs. Figure 2.5(a) shows a simplified FSMD with 2 states Si and Sj and two transitions. Each transition is defined with a condition and a set of actions that have to be executed in parallel when the transition is fixed. Figure 2.5(b) shows an implementation of the FSMD using a structure made of a controller and a data path. The controller implements the FSM. It computes the next state the control output and the data path control line according to control input lines, status lines and the present state. The data path is made of a set of storage units and a set of calculation units.

The FSM with co-processors model (FSMC)

An FSMC is an FSMD with operations executed on co-processors. The expressions may include complex operations executed on specific calculation units called co-processors. An FMSC is defined as an FSMD plus a set of N co-processors C. Each co-processor Ci is also defined as an FSMC. FSMC models hierarchical architecture made of a top controller and a set of data path that may include FSMC components as shown in Figure 2.6.

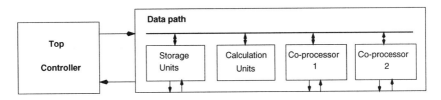

Figure 2.6 FSMC Architecture Model

Co-processors may be as complex as FSMCs. They have their local controller, inputs and outputs. They are used by the top controllers to execute specific operations (expressions of the behavioral description).

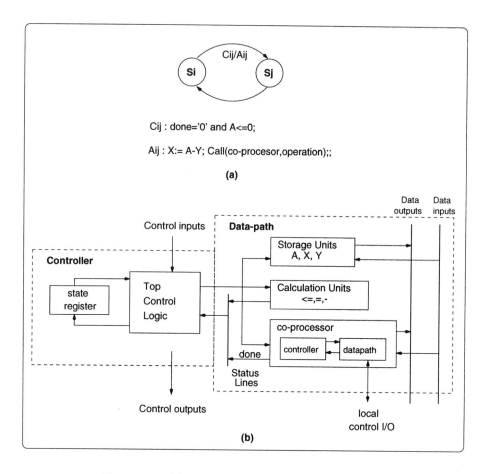

Figure 2.7 (a)FSMC Model, (b) FSMC implementation

Definition 2.6 (FSMC) *The FSMC may be formulated as :*

$$< S \times \Pi_{i=1}^{N} S_{ci}, I \times SS \times \Pi_{i=1}^{N} I_{ci}, O \times A \times \Pi_{i=1}^{N} O_{ci}, f, h >, \text{ Where :}$$

- S_{ci} : *The set of states of co-processor i*
- I_{ci} : *The set of inputs of CO-processor i*
- O_{ci} : *The set of outputs of co-processor i*

Models for behavioral synthesis 33

- $S \times \Pi_{i=1}^{N} S_{ci}$: *The set of states of the FSMC. The set of the FSMC is defined as the product of the state of the FSM and the local states of the co-processors.*

- $I \times SS \times \Pi_{i=1}^{N} I_{ci}$: *The inputs of the FSMC. Inputs are extended with status (SS) and local inputs of co-processors.*

- $O \times A \times \Pi_{i=1}^{N} O_{ci}$: *The outputs of the FSMC. Outputs are extended with assignments and local outputs of co-processors.*

The behavioral expressions translating specific operations may include operations that are not standard in the language employed to describe the behavior of a given specification (VHDL for instance); these specific operations may be represented as procedure and function that will be executed on a co-processor. Figure 2.7(b) shows an implementation of an FSMC which is also made of a controller and a data path. The controller executes the top control. It computes the next state, the output and the control line of the top controller according to control input lines, status lines and the present state of the top controller. The data path is made of a set of storage units, a set of calculation units and a set of co-processors. The co-processors may have their internal state and I/O. This model is used by the AMICAL system. It will be detailed in chapter 5.

2.2 THE DATAPATH CONTROLLER MODEL

2.2.1 Global Organization

Design of processors mostly follows the well known scheme of a von Newman architecture. This universal model of computing systems can be partitioned into two parts : an **operative part**, also called execution part or datapath, and a **control part** or controller. All these notations will be employed equally in this book. This model is commonly used for the design and the synthesis of microprocessors, which are reusable devices [Anc86] targeting software applications. This model is widely used as well for behavioral synthesis [GDWL92, HT83, VRB+93, MLD92, Inc94], which targets very specific and dedicated embedded computing systems.

As its name indicates, this architecture consists of two parts: a controller and a datapath. Roughly, the datapath is the place where things happen, while

34 CHAPTER 2

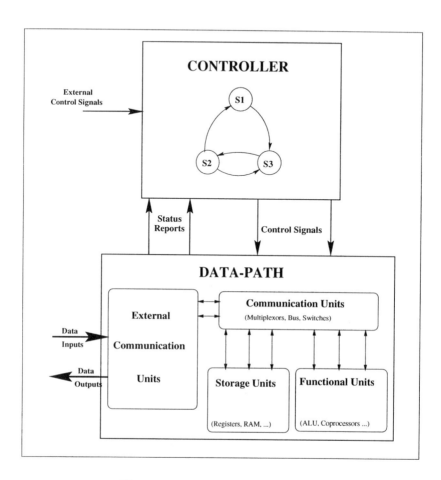

Figure 2.8 General Architecture Model

the controller is the place where decisions are taken. Figure 2.8 depicts the main features of both parts and their interconnection. The controller or control part consists of a finite state machine (FSM) which inputs the external control signals and datapath flags (execution or status reports), and drives the operative part through its control signals. The datapath or operative part is the execution part. It is composed of four types of components or units :

1. **functional units** : also called operators or computation units in the literature; execute operations specified in the behavioral description; this type

Models for behavioral synthesis 35

of unit may be simple, such as adder, multiplier, shifter, or more complex such as co-processors including local control.

2. **storage units** : hold the values of variables and constants generated and consumed during the execution of the behavior : registers, register files, RAMs and ROMs.

3. **communication units** : construct the communication network for data transfers between storage units, functional units and external ports : Buses, multiplexors, switches.

4. **external communication units** : realize the interfaces between external ports and the other units of the datapath.

This architecture is synchronous, i.e a change of internal state, followed by actions within controller and datapath, occurs on clock signal events. We will consider a single phase clock.

2.2.2 Synchronization models

Synchronization involves the clear definition of what is a clock cycle and what happens during it, i.e what are the actions involved in control and operative parts and how these actions will be scheduled between both parts during it. In other words, we need to know the timing scheme, or synchronization model. The synchronization model of a circuit made up of a datapath and a controller defines the synchronization instants of a basic cycle according to the clock of the circuit , i.e defines how control and execution parts must act and converse during the cycle. Its definition will have large consequences on the result of the synthesis and therefore on the final architecture.

Basic execution cycle, clocks

Globally, one can distinguish four basic steps which are executed successively each time the system, composed of both control and operative parts, operates.

1. The controller enters a new state, computes the control signals and the next state;

2. The controller sends the control signals to the datapath;

3. The datapath executes the operations corresponding to the controller control signals;

4. The result of the datapath transfers are stored, and the status lines, if required, are sent to the controller.

Control signals and datapath status lines may or may not be latched, according to the existence of a control pipelining, as we will see further. We denote by $t1$, $t2$, $t3$ and $t4$ respectively the execution time associated to the critical path of the controller, the control signals circuitry, the datapath and the status signals output circuitry. These successive instants are depicted on figure 2.9.

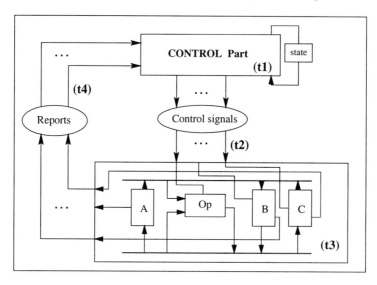

Figure 2.9 Schedule of events inside a circuit.

The way these time slots will be organized within a clock cycle will define the synchronization between control part and datapath, and then will impose different architectural styles. Two synchronization schemes are described in the following subsections, whose difference resides in the existence of a pipeline stage or not between control and operative part.

Classic model : non-pipelined architecture

In the classic model, a clock cycle must include the execution of the four different time steps defined above. Consequently, the total duration d has to be the concatenation of the four time slots defined above :

$$d = t1 + t2 + t3 + t4$$

In general, time slot t2 and t4 are negligible. So the total duration of the global computation time within a clock cycle is approximately equal to :

$$d = t1 + t3$$

This simple clocking scheme is the one without overlapping : during one clock cycle, both controller and datapath execute a fixed set of operations sequentially. With this scheme, we must be careful when synchronizing both subsystems. When the controller uses the flags to evaluate a set of conditions, it must be sure that the appropriate condition signals have already been generated by the datapath. In other words, we must add restrictions on the timing to ensure that the controller and the datapath work well together in the same basic cycle; we then must have, if T is the clock period :

$$T \geq d.$$

Figure 2.10 illustrates this clocking scheme. On the rising edge of the clock, the controller enters a new state, computes the control signals, and sends them to the datapath. The emitted control signals are supposed to reach the datapath within the time slot, t1. The datapath performs the appropriate operations during the time slot, t3, and updates the registers on the next rising edge of the clock.

This model presents the disadvantage of imposing a clock that must be slow enough to cover the critical path execution time of both the controller and the datapath.

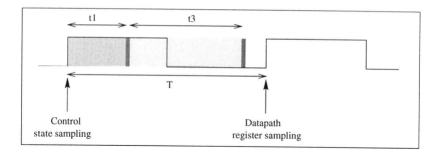

Figure 2.10 A simple clocking scheme.

Pipelined architecture : control pipelining

Another approach would be to pipeline the execution of the controller and the datapath. This corresponds to one of the three available pipelining possibilities introduced in chapter 1, and is called **control pipelining**. In the corresponding timing scheme, the datapath executes transfers corresponding to the control signals generated at least during the previous clock cycle. In other words, controller and operative part work in parallel on consecutive instructions, as depicted on figure 2.11. In order to allow such a timing configuration, two constraints have to be respected :

1. The control signals and/or the reports signals have to be stored in additional registers (figure 2.11.a).

2. The transition conditions should not depend on the results of the current state operation but on the previous one.

If the generation of the command vector for cycle $i+1$ depends on data calculated in the datapath for cycle i, this necessitates the insertion of idle cycles during the execution. Such a configuration is depicted on figure 2.11.c, and for a pipeline latency equal to 1 (control part fixes the control vector for the operative part, for the clock cycle following immediately). This timing scheme, if chosen to be targeted by a synthesis system, will therefore increase the synthesis algorithms complexity. Indeed, in this case the scheduler, which fixes the scheduling of operations during the synthesis process (as will be tackled in chapter 3), must introduce judiciously these idle cycles, and therefore, must restructure the way the actions are scheduled. Roughly, when a pipeline dependency is found, the scheduler inserts an idle cycle in order to allow the

Models for behavioral synthesis

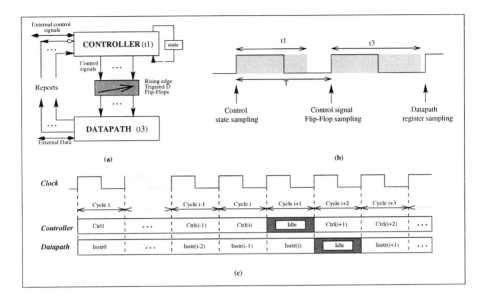

Figure 2.11 A clocking scheme with a controller pipelining.

controller to read the status sent by the datapath, and to evaluate the right conditions. This is explained on figure 2.12 : a succession of actions A depending on conditions C (C/A) are depicted, where the evaluation of the condition $Ci + 1$ depends on data calculated in the action Ai from state Si. If a control pipeline of latency 1 is targeted, it is then obvious that it will be not possible to calculate condition $Ci + 1$ in parallel with the execution of Ai, because $Ci + 1$ will be able to be calculated only at the end of Ai. This is why an idle cycle must be inserted between the states Si and $Si + 1$ (insertion of a new state), so that a control pipelining becomes applicable on this scheduling with success. We will then find the configuration of figure 2.11.c.

This synchronization model will also increase the global amount of area needed. Indeed, for instance in the case of a pipeline depth of one as seen above, a set of additional registers will be inserted between control and operative parts in order to store the intermediary control vector value (the reports signal being already stored in the datapath registers, we don't have to store them twice). However, the choice of this timing scheme will allow to speed up the circuit in some particular cases. Indeed, in this synchronization model, the clock period must follow :

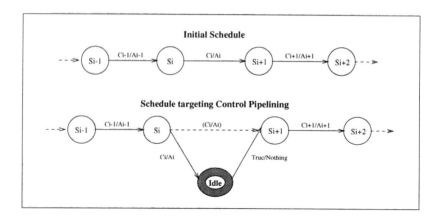

Figure 2.12 Introduction of an Idle cycle.

$$T \geq Max(t1, t3)$$

which is $< d = t1 + t3$, if we consider the previous synchronization instant definitions (figure 2.11.b). Thus, in the case of a non intensive use of report signal from the datapath, e.g in the case of a weak data dependent control behavior, a few amount of idle cycle will be inserted, and then the gain in speed will be consequent. In other words, it is clear that the choice of the clocking scheme must be done early, because it conditions all synthesis steps. This choice must be based on a good knowledge of how the specification works : for instance, if the control often waits for action reports, which means insertion of a lot of idle states, the pipelined model may be of little interests.

2.3 DATAPATH MODELS

Conceptually, a datapath may be defined by two aspects, which resume the possible architecture styles. First, the internal structure, including all types of units allowed, and second the functional organization, i.e the types of transfers allowed, how these transfers are executed and how communication with the controller is performed in both directions. Each of these parts will be developed and described in a generic way in the following sections.

2.3.1 Datapath structure

As introduced previously, a datapath architecture is defined as a set of components associated to an interconnection topology. For behavioral synthesis, a simple target architecture may greatly reduce the complexity of the synthesis problems. On the other hand, a less constrained architecture, although difficult to synthesize, may result in higher quality designs [GDWL92, VRB+93, MLD92].

Datapath representation

The basic conceptual architecture organization is depicted on the figure 2.13.

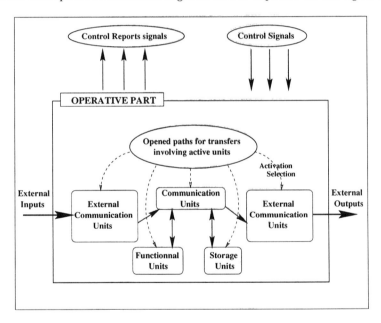

Figure 2.13 Conceptual organization of the data-path architecture

Data exchanges between units are done through the use of communication units. Transfers follow the paths opened by the control signals. They obey to a transfer model which includes the instruction set, defining the dependencies between source and destination for a transfer, and the generation rules which define the realization of a transfer. This point will be tackled in a further section.

Each unit within the datapath is more or less built on the generic scheme represented on figure 2.14 ([DR85]).

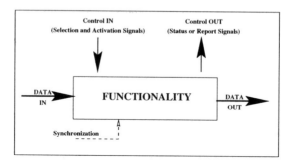

Figure 2.14 Generic model of a datapath unit

Roughly, each unit manipulates data and transforms, transmits or stores these data. Therefore, each type of unit will differ from the others principally by its functionality, while the connection with the surrounding components will be done following similar schemes.

Functional units

A functional unit is a specific datapath unit aimed to execute the operations of the behavioral description. The term functional unit or FU used here is very general and means all types of units executing transformations on data. This include the execution units (EXU) and application specific execution units (ASU) as defined in [VRB+93]. This also includes co-processors as defined in the FSMC mode.

a. Structural organization

A FU may execute one or several operations. $OP1, OP2...OPn$, as depicted on figure 2.15, represent possible operations performed by the component, whose execution depends on selection signals. These operations may be of different types :

- **logical operations** : AND, OR, Exclusive OR, NOT, ...;
- **arithmetic operations** : +, -, increment, comparison ...;

Models for behavioral synthesis 43

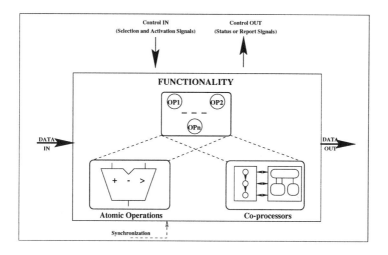

Figure 2.15 Functional unit representation

- **shift operations**;

- **complex operations** : may correspond to predefined operations in the behavioral description, such as procedure and function calls for instance, which in turn may execute on a co-processor (see chapter 4).

Each operation is characterized by an execution time expressed in number of clock cycle. In order to perform the synthesis steps, the behavioral synthesis system must have accesses to the timing information of each unit involved in the synthesis process. This information will conditions important decisions during synthesis, such as scheduling and choice of the optimal clock period.

b. Component pipelining

The goal in component pipelining is to increase the unit utilization, and eventually shorten the clock cycle, which results in an increased throughput. Components are divided into two or more stages, while the result of each stage is latched before reaching the next stage. Let us examine a simple example : a multiply-accumulate FU, widely used in DSP applications (figure 2.16).

Let us suppose that a long stream of inputs is applied each clock rising-edge. In figure 2.16.a, the initial component performs operations + and × sequentially

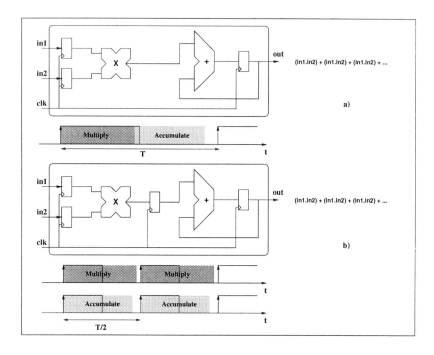

Figure 2.16 Example of component pipelining

in the same clock period T. T, as illustrated, must be at least equal to the sum of the critical path execution time of both operations. Moreover, as they execute sequentially, each unit remain idle while the other is active, which results in an under-utilization. This functioning can be improved by latching the result of the multiplier. In that case, the final component presents the following improvements :

- The clock period has been shortened, at least equal to the maximum critical path execution time of both components, which is anyway less than the previous delay.

- Both operations work in parallel, after a latency of two clock cycles, which leads to a throughput increased by a factor of 2, as the output port gives a new result each half period T.

The main drawback of such a pipelining is the resulting latency, which is defined as the delay that a circuit takes to process a given set of input data and produce

Models for behavioral synthesis 45

the corresponding outputs. For instance here, the circuit produces outputs two clock periods after the introduction of the corresponding inputs : it corresponds to a latency of 2; but data succeed on the input ports each clock cycle. If allocated by a synthesis system, the component latency will be part of the information available for it, which will especially influence the scheduling step.

Storage units

The storage units (SUs), such as registers, register files, RAMs and ROMs, are the components which can store the variable values or provide constants specified in the initial behavioral description. As they manipulate data through a protocol that may be complex, register files, ROMs and RAMs can be described as special functional units which shall execute the storage functions through two operations called for instance "read" and "write".

A basic set of SUs may be the one described on figure 2.17.

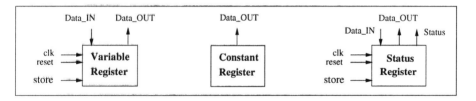

Figure 2.17 Basic SU set

- **A Constant register** is used to generate a single constant; useful if there is a few number of constant needed (otherwise, a ROM is recommended);

- **A Variable register** is a classic register which allows to store new values within it, and otherwise maintain its value on its output port; the signal "store" determine in which mode the register has to be set;

- **A Status register** looks like a variable register, but has another output, called here **status**. This kind of register is needed when the controller makes use of values generated by the datapath to evaluate the conditions (i.e generally when comparison units are placed within the controller itself).

We assume here that the value of the SU is always available on the Data-Out port. Of course, other access model are possible.

Communication units

The communication units (CU) or transmission units are used to control data transfers between other units, through nets and buses which are the supports of these data transfers (see section 2.3.2). Figure 2.18.a depicts the generic form of a transmission unit.

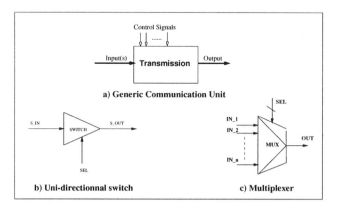

Figure 2.18 Basic communication units structure

There are mainly two types of communication units (figures 2.18.b and 2.18.c):

- **Switches** : uni or bi-directional, the switches transfer data linearly respectively in one or both senses;

- **Multiplexers** : the multiplexers (MUXs) control the data transfer from multiple inputs towards one output;

The switch acts as a tri-state driver, whose output may be connected to a bus. This very simple model can be implemented with only few transistors. A multiplexer (mux) allows to select one input from N. Generally, $N = 2^M$ were M is the bitwidth of the select signal.

Data transfers between datapath components make use of a communication network which is made itself of communication units and their supports or wires (net and buses). The choice of a type of communication units fixes the topology of the communication network :

Models for behavioral synthesis

- **shared-interconnection topology** : uses global buses; this means that the same bus can be shared by several data transfers, so buses are shared-interconnections; this first method maximizes resource sharing for interconnection. Some buses may even be cut into segments so that each bus segment is used by parallel data transfers. A data transfer between the segments is achieved through switches between bus segments. This scheme requires sophisticated algorithms to be included in the behavioral synthesis process.

- **point-to-point topology** : this method simplifies the connection allocation algorithms : a connection is created between any two units if needed; if more than one connection is assigned to the input of a unit, a multiplexer is used.

Buses are multi-sources nets. They can be seen as specific units, because of their possible intrinsic complexity, or as a set of single net. A bus being connected to several sources, their access must be carefully protected in order to avoid conflicts; this may be done through the use of switches, and requires a careful design. Connection allocation is an important step during behavioral synthesis. It may influence the resulting performances. This is generally performed according to a trade-off speed/area as illustrated by the following example (figure 2.19)

On figure 2.19.a, a single bus is used to realize the transfers between registers A and B, and the binary (two operands) operator. Uni-directional switches 1, 2 and 3 are placed on each data source (unit output) in order to be able to control their write on the bus and then to avoid conflicts, i.e several simultaneous write on the bus. Inputs of FU are latched by a and b in order to keep the values transmitted before starting to compute. We assume that write into each register is enabled or not by a **store** control port, and occurs on a rising-edge of the global clock. Obviously, the non-used switches in a given cycle are maintained off by control signals, while the useful switches are kept on. In this configuration, 3 cycles are required to execute operation $A <= FU(A, B)$, through three distinct transfers (see 2.3.2) :

1) $A \rightarrow FU[a]$ *through* 1
2) $B \rightarrow FU[b]$ *through* 3
3) $FU(s) \rightarrow A$ *through* 2

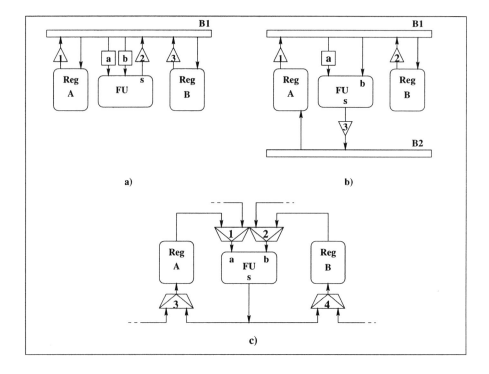

Figure 2.19 Example of communication through switches and MUXs

Figure 2.19.b shows a double bus configuration which allows to execute the same operation in two cycles only, the first transfer being performed during the first cycle through 1, and both last data exchanges during the second cycle, in only one transfer, the result of the operation being stored in A at the end of the cycle (on the rising-edge of the next clock cycle) through bus B2. In this case, only one FU input need to be latched.

This basic operation can even be executed in only one cycle if a third bus is added, allowing storage of the result in register A at the end of the same operating cycle (on the rising-edge of the next clock cycle). In this last case, neither latches nor switches are necessary. The same result is obtained on figure 2.19.c, where four multiplexers are used instead of buses and switches (we assume that the other multiplexer input ports are bound to other unit output port, not represented here).

External communication units

The external communication units (ECUs) link the operative part with the external world. These elements may be very simple, only composed of buffers; however, they can also be used to carry out simple transformations such as data type conversion, modification of the number of bits as well as complex transformations such as AD/DA conversion.

Figure 2.20 shows several basic ECUs. The output depends on the inputs and on the selection signals.

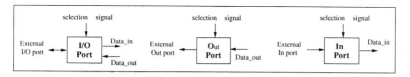

Figure 2.20 Basic set of external communication units

2.3.2 Functional organization

A datapath, as we have already seen, is the execution part of the whole circuit. At each basic clock cycle, it executes an instruction, itself possibly composed of a set of parallel transfers. The computation power of a datapath can be fixed by the transfer model associated to it, and by the amount of parallelism it allows. The "instruction set" may be defined by the set of transfers that can be executed by the datapath. Therefore, in addition to the structural information, the datapath definition must includes a **transfer model**.

General data transfer model

a. Transfer definition

A transfer is defined by a source, a sink or destination and a path. Both source and destination are components of the datapath. During a transfer, data are either exchanged, or transformed, or both, sequentially, and this during one basic clock cycle.

As depicted on figure 2.21, this definition involves a transfer path between source and destination units; a path can be composed of others datapath units such as CUs, but also FUs. A transfer can therefore be represented as a graph,

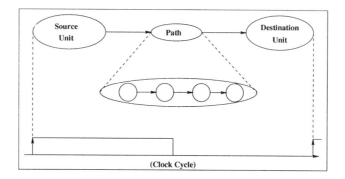

Figure 2.21 General transfer

where nodes are the datapath components, and the edges are the physical links between them. Each transfer may then be decomposed into several atomic data exchanges from the source unit to the destination unit. In order to fix how distinguish source and destination units from path units, one can define the source unit as the first unit involved after the beginning of the clock cycle, and the destination unit as the last unit reached before the end of the clock cycle. If we consider the previous component sets, the general form of a transfer is the following :

	TRANSFER	
$Source$ $\Rightarrow$	$Path$ $\Rightarrow$	$Destination$
[**ECU, SU,** FU] $\rightarrow$	$CU \rightarrow \{FU, CU\} \rightarrow CU$ $\rightarrow$	[**ECU, SU,** FU]

b. A particular transfer type : chaining

Transfers that include several FUs such as $FU \rightarrow CU \rightarrow FU$ are particular : they correspond to the term **chaining** in behavioral synthesis, i.e chain two operations or more during one cycle. Figure 2.22 gives a simple example of chaining and how useful it can be.

Figure 2.22.a depicts a Data Flow Graph representing a simple operation : $e = a + b + d$, decomposed into two operations $c = a + b$ and $e = c + d$ using binary adders. Let us assume that the clock cycle has been definitively fixed and that the adder has a maximum delay less than half the clock period (we assume that the multiplexers delay are negligible). On 2.22.b, the execution of

Models for behavioral synthesis

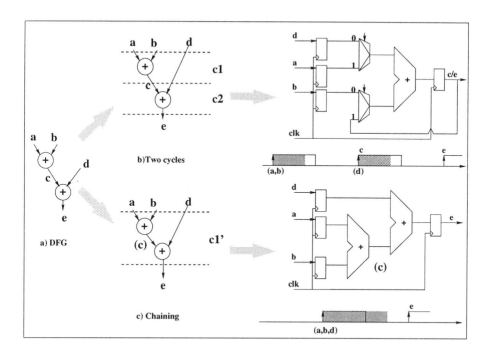

Figure 2.22 Example of particular transfer : chaining

these successive operations is cut into two clock cycles, as the corresponding sequencing graph shows. The transfers involved in both cycles are of the same type that on figure 2.19.c in section 2.17. Intermediate value of variable c is stored between both cycles (in a register). We can immediately note that a large part of the clock cycle is useless, which is not optimal. Figure 2.22.c shows a chaining of both basic addition operations into one cycle. The global delay involved remains less than one clock period : the resulting circuit works faster, as the global time cost is reduced to perform the global operation.

Parallel transfers and datapath pipelining

a. Notions of parallel transfers

In the previous section we defined the notion of transfer : sequential exchange and/or transformation of data, and this during one clock cycle. However, several transfers may occur at the same time, i.e within the same clock cycle : **parallel transfers**. This allows the execution of several operations simultaneously during a given cycle. Instructions, defined as the set of transfers executed within the datapath during a given clock cycle, may concern a lot of units. The main goal of parallelization is to increase the performance in term of number of execution cycles and in term of global utilization. A simple example has already been studied on figure 2.19.c in section 2.17 : operation A = FU (A,B) is performed in just one clock cycle. It is possible because transfers of the content of the registers towards the FU inputs are done in parallel : part one of the global instruction is composed of two "atomic" data exchanges executed simultaneously.

$$\left[\begin{array}{cc} (1) & (2) \\ \left. \begin{array}{l} Reg_A \rightarrow Mux_1 \rightarrow FU_a \\ Reg_B \rightarrow Mux_2 \rightarrow FU_b \end{array} \right\} & FU_s \rightarrow Mux_3 \rightarrow Reg_A \end{array} \right]$$

But the concurrency may involve several arithmetic operations, for instance as in the case of a micro-processor where the datapath may be split into for instance address calculation part, and data processing part, both working together ([Anc86]). Figure 2.23 gives an example of operations executed concurrently.

Fig. 2.23.a shows a DFG, scheduled on fig 2.23.b. In this case, cycle 2 contains two parallel operations (+). As illustrated on figure 2.23.c, the implementation allows the execution of both addition operations in parallel. The communication between registers and FUs is ensured by the use of MUXs.

b. Datapath pipelining

An other architectural aspect that conditions concurrency of transfers is the pipelining of the whole operative part, aimed to increase utilization and computation speed similarly to the component pipelining. Mostly applied for signal processing applications, which generally present regular data stream feeding a datapath performing the same sequence of operations on each sample of the steam, datapath pipelining consist in splitting the whole datapath into two or more stages. Two consecutive stages are separated by registers in order to store the intermediary results, so that each stage can work on a sample of the stream concurrently with others (working on previous or next sample) and provide a result for the next stage, and so on. Each stage may execute in more than one clock cycle; in that case, all stages must have the same local latency in order to work together in good conditions and update input data at the right moments.

Models for behavioral synthesis 53

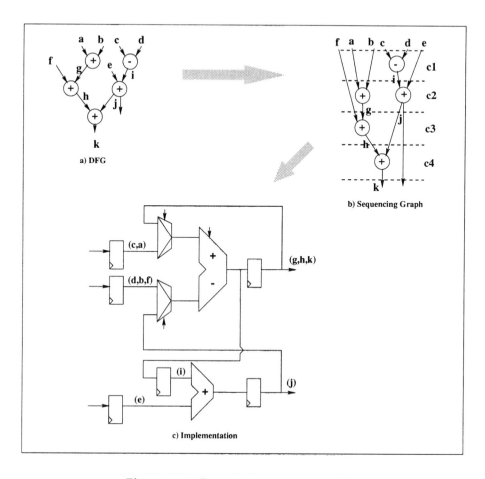

Figure 2.23 Example of concurrent operations

Let us assume a non-pipelined datapath fed by a regular data stream, and performing a sequence of operations on each sample with a global latency of 9 cycles, as depicted schematically on figure 2.24.a. With this initial configuration, a new sample of data is treated every 9 cycles. In order to increase the input/output data rate or **throughput**, a 3 stage pipelining is applied, each stages having a pipeline latency of 3 cycles (figure 2.24.b). The global latency is not modified, as 9 clock cycles separate a data sample input from its corresponding output. However, this new configuration allows a new data sample to be introduced each 3 cycles, which leads to a throughput being increased by a factor of 3, as a new processed data sample get out every 3 cycles instead of every 9. Actually, it is possible because all stages work in parallel (after the

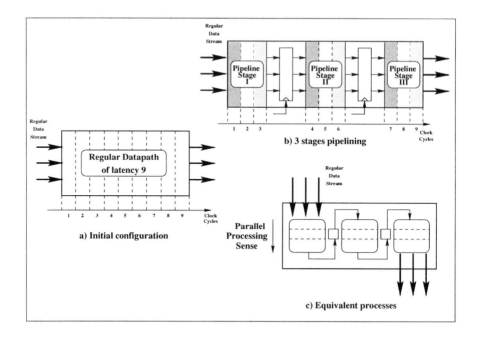

Figure 2.24 Datapath pipelining scheme

first 9 cycles following the introduction of the first sample); indeed, time steps $(1, 4, 7)$, $(2, 5, 8)$ and $(3, 6, 9)$ work concurrently. This is just as if the initial datapath had been cut into three specific processor, each one corresponding to a pipelining stage, with fixed processing delay (figure 2.24.c). At each clock cycle, the instruction sent by the control part opens the paths for three parallel transfers. The introduction of pipeline stages within an operative part can also lead to a reduced clock period, if it permit to reduce the transfer length i.e the critical paths (see component pipelining).

2.3.3 Examples of datapath target architecture

Several datapath architectures suitable for behavioral synthesis are described in this section. These correspond to restricted versions of the generic model defined above. For instance, the transfer model employed here corresponds to a sub-set of the one described previously :

Models for behavioral synthesis

- Direct transfers (through CUs) between FUs are forbidden. In other words, this transfer model don't allow the chaining of two FUs.
- Direct transfers between ECUs is not allowed : no data transmission from an input port towards an output port without storage or transformation.

Two datapath target interconnection topology are tackled here : a bus-based solution, and a multiplexer-based solution.

Bus based datapath

In this model, the internal connections or interconnections between units are made up of buses and uni-directional switches.

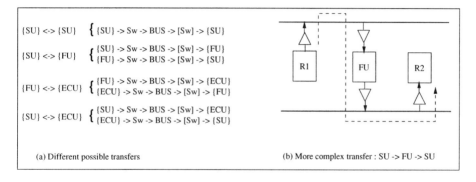

Figure 2.25 Transfers execution for a bus-based architecture

The execution of the basic transfers allowed by the transfer model and mapped to the bus model is detailed on figure 2.25. Sw and $[Sw]$ mean respectively switch and optional switch.

More complex transfers can be realized by the combination of several basic transfers (figure 2.25.b), as already seen in the previous examples. Figure 2.26 shows a bus-based datapath example generated for the GCD (Greatest Common Divider) whose VHDL behavioral description is given on part a). The corresponding datapath is made up of 2 registers(x and y), which both correspond to the variables of the description, one functional unit which realizes operation $(-)$ and four external ports (x_i, y_i, $output$ and $dout$). Ports $start$ and din are connected to the control part. As the controller needs to compare the values of x and y (comparators are grouped in the control part, which is another important design decision, as will be seen further), status-registers are

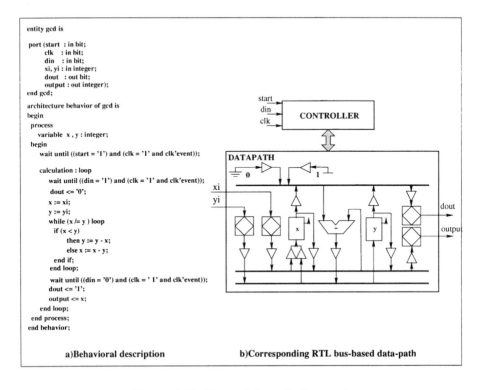

Figure 2.26 Bus model: synthesis example

employed instead of simple storage units. The datapath includes also 13 unidirectional switches and three buses : this allows to perform operations such as $y <= y - x$, or $y <= yi$ and $x <= xi$ in only one cycle, in a single instruction. This circuit is not optimal in term of area. For instance, both switches on the input of register x could be advantageously replaced by a single mux.

Multiplexer based datapath

With this model, we can directly transfer data between the external ports and registers or functional units through only one multiplexer.

The transfer model is the same as in the previous section. The execution model for these transfers are detailed as depicted on figure 2.27. Compared with the bus model, the transfer model lead to similar transfer types. But the transfer executions are simpler : less units are used to realize a transfer. It means that

Models for behavioral synthesis

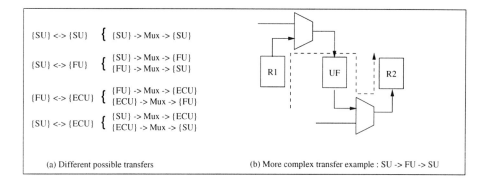

Figure 2.27   Transfers executions for a mux-based architecture

less control signal will be required; it means also that the transfer delays are smaller.

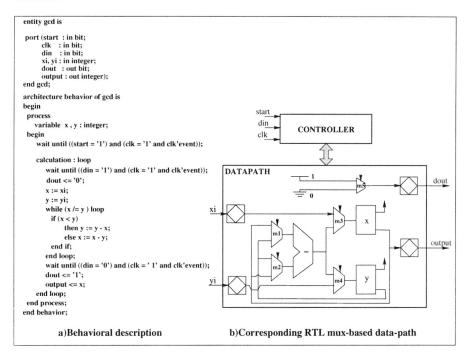

Figure 2.28   Multiplexer model: synthesis result

Figure 2.28 presents a mux-based datapath for the same GCD example. Here, five MUXs are used. The resulting architecture is much simpler than in the bus-based one. Connections are realized directly between units.

2.4 CONTROLLER MODELS

The control part of a general circuit may be seen as the decision center : it orchestrates all transfers within the datapath. This section focus only on the general organization of controllers at the RT level, and on several possible implementations suitable for behavioral synthesis. There are two main styles used for the physical implementation of a controller : the flat FSM style, which directly translates the behavior into random logic or PLA, and the micro-coded styles which makes use of a micro-coded ROM generating the wanted control signals with the help of a sequencer.

2.4.1 Theory and general organization of controllers

A controller based on a Finite-State Machine (FSM) is the most popular design model. In this section, we will use the FSM theory to define the controller representation[GDWL92]. A sequential system such as a FSM can be represented as depicted on figure 2.29, decomposed into a combinational part and a state storage part which is synchronized in the case of a synchronous sequential system. The model consists of a set of states, a set of transitions, and a set of actions. These last may be associated with states, transitions or both states and transitions. More formally, a FSM can be described as a quintuple:

$$FSM = (S, I, O, FS, FO)$$

where :

Models for behavioral synthesis

$S = \{s_i\}$ is the set of states;
$I = \{i_j\}$ is the set of input ports;
$O = \{o_l\}$ is the set of output ports;
$FS = \{fs \mid fs : S \times I \rightarrow S\}$ is the set of next-state functions that map a cross product of S and I into S (sequencing);
FO is the set of functions updating the output ports (command generation).

In other words, FS is the set of functions giving the next state considering a present state and the input ports values. Two main theories conflict on the definition of FO : the first one leads to FSMs called MOORE Automaton, and the second one to FSMs called MEALY Automaton.

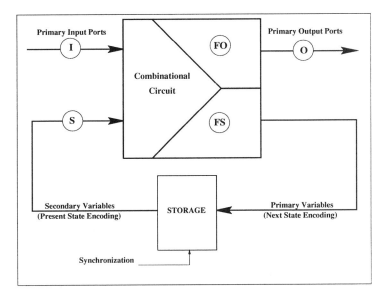

Figure 2.29 General Synchronous Sequential System Organization

In the case of a MOORE Automaton, FO can be defined as $FO = \{fo \mid fo : S \rightarrow O\}$, while in the case of a MEALY Automaton, it is defined as $FO = \{fo \mid fo : S \times I \rightarrow O\}$. In other words, in a MOORE Automaton, outputs calculation only depends on the present state, while in a MEALY Automaton, outputs calculation depends not only on the present state, but also on the input values. The Moore approach is also called **conditional sequencing**, while the Mealy approach is called **conditional generation of command** ([Anc86]). Primary and secondary variables are respectively the next and present state lines. The number of states fixes the number of state lines and then the number of storage elements needed. N states will require a number m of storage elements equal to :

$$m = INT[log_2(N)]$$

Sequencing

The sequencing of an FSM, either of Mealy or Moore type, is performed through the use of synchronized storage elements, whose choice is critical for a correct functioning ([Anc86]). Indeed, latches are not recommended because of their transparent mode : during it, the FSM becomes an asynchronous automaton, whose internal states can change unpredictably. Edge-triggered flip-flops may be dangerous as well, although their use is less critical than latches. Indeed, as nothing is perfect, the loading of each state line never occurs simultaneously, which can lead to transitory and useless computations and then to transitory unwanted next-states and outputs. The best solution, which solves all these problems, is to employ a **Master-Slave** structure. This structure consists of two storage element layers, driven by non-overlapping phases. Others sequencing solutions are employed, which consist for instance in decomposing the combinational network into several blocks operating alternately, and corresponding to two or more non-overlapping phases. We will focus here on single phase controller implementation. If we consider the case of a Mealy Automaton, when certain conditions are satisfied, and considering a given state, a next state is fixed and of course a transition. If this transition is associated to output assignments, then paths will be opened in the datapath during this transition. Therefore, when the FSM enters a new state, generally on a clock edge, next-state is calculated and outputs to the datapath are calculated and updated or stored, depending on the synchronization scheme (see section 2.2.2). An example of sequencing is depicted on figure 2.30, where a clock cycle is roughly cut into two parts, as introduced in section 2.2.2 : the first part concerns the controller, and the second is reserved to the datapath.

The main difference between Mealy and Moore machine is that output assignments are fixed in a Moore FSM because they only depend on the present state, while they are not in a Mealy FSM : if an input value changes during a control cycle, i.e between two clock edges, then output assignments and next state may be modified consequently.

Models for behavioral synthesis 61

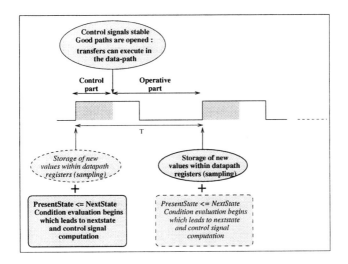

Figure 2.30 Sequencing of the gcd example

2.4.2 Controllers organization

Hardwired control model : the flat FSM controller

There are several ways to physically implement a controller. One way is to describe it as a flat FSM, using Mealy or Moore organization, and then synthesize it into gates or PLA. It supposes a way to describe the control behavior in a synthetizable language. The following examples present two ways, graphical and language (VHDL), to describe a given behavior in both organization : Moore and Mealy. The state graph of a simple Moore FSM is depicted on figure 2.31.

A state graph is a graph where nodes represent states and arcs represent transitions between states. Arcs are generally associated with conditions. For instance, on figure 2.31, the transition from $S0$ to $S2$ is executed if input X is equal to 1, otherwise the machine stays in state $S0$. We can see that an output value of Z is associated to each state. The state graph of a Mealy FSM is depicted on figure 2.32.

The difference between both graphical description is obvious : in the case of Mealy automaton, output values are associated with conditions on input values; in other words, instead of being associated with states, output values

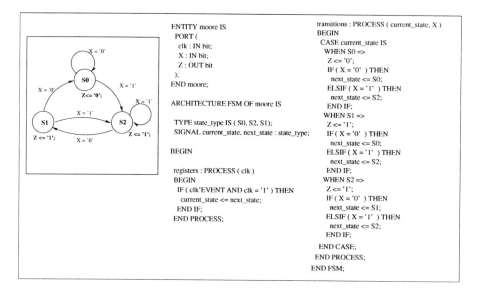

Figure 2.31 Moore Automaton : VHDL Description

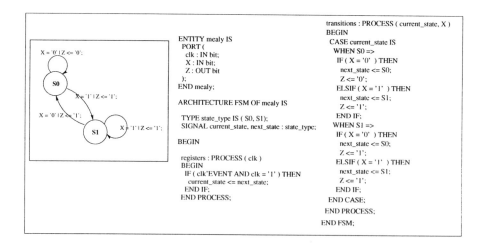

Figure 2.32 Mealy Automaton : VHDL Description

are associated with transitions. This difference is easily mapped to the VHDL description; if we compare it with the previous one, we can see that output assignments are now associated with conditions (if ... then) instead of states (when S1 ...). Both described machines present the same behavior : when in $S0$,

Models for behavioral synthesis 63

i.e the initial state, while $X = 0$, $Z = 0$; if X takes the value 1, $Z = 1$ and keep 1 while $X = 1$; then, if X takes twice the value 0 consecutively, Z takes value 0 again, else keep value 1. However, one can remark that the Mealy description presents one state less than the Moore description. In general, a Mealy machine can be easily converted to a Moore machine by adding additional internal states. For that reason, it will be sometimes better to choose a Mealy approach, as the cost in term of area and performances of a sequential machine highly depends on its number of states .

Programmable control model : micro-coded controller

In this implementation style, the controller is a programmable micro-controller. This type of architecture presents advantages such as reusability, design flexibility (mistakes can be more easily corrected), advantages which can be useful during a system design, the main drawbacks being the loss in performance (area and timing).

Several programmable models for controllers are reported in the literature [Anc86]. The general architecture is depicted on figure 2.33.

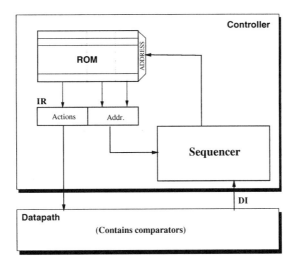

Figure 2.33 Micro-programmable controller architecture

This general model is composed of three main blocks :

- a ROM containing the control code or control vectors or micro-code;
- an Instruction Register (IR) containing the control signals, the next instruction address and next address calculation mode flag. This last part indicates if intermediary states must be inserted in case of data dependencies between two consecutive instructions.
- a sequencer giving the next address knowing the content of the IR and the processing results from the datapath.

We will see in chapter 3 what kind of difficulties have to be handled with this target model, particularly for the mapping of a controller onto such an architecture and the necessity of rescheduling associated.

2.5 SUMMARY

The intermediate form and the target architecture are the two main models used during the behavioral synthesis.

The intermediate form fixes the underlying design model of the behavioral synthesis tool. The synthesis process is generally decomposed into a set of refinement steps acting at the behavioral level. Each of these steps is described through the use of an intermediate form, which can be language oriented or architecture oriented. Language oriented intermediate forms are based on graph representations (CDG, DFG, CDFG). Architecture oriented forms are based on FSM models. Both kinds of intermediate forms may be combined during the behavioral synthesis process.

The target architecture of most existing behavioral synthesis tools is made of a datapath and a controller. Several styles are defined within this generic model.

The global synchronization of the circuit involves the clear definition of what is a clock cycle and what happens during it. In this context, a pipelined communication between the controller and the datapath, called control pipelining, may greatly influence the way the global synchronization is tackled. The datapath is defined by its internal structure and its functional organization. The former fixes the set of components used by the datapath and the communication styles within it (multiplexer or bus based). The latter defines the instruction set, which itself can be defined by the basic transfer model and the parallelism offered by the internal structure. The controller organization depends on its description style (Moore or Mealy), and on the way it is physically implemented (flat or micro-coded FSM).

3
VHDL MODELING FOR BEHAVIORAL SYNTHESIS

The most difficult task, when using a behavioral synthesis tool, is to predict the resulting architecture while writing the behavioral model. This comes from the fact that the behavioral synthesis process involves complex transformations of the behavioral description that may induce a large reorganization of the initial model. Besides, each tool imposes specific restrictions and may use different algorithms.

This chapter deals with three issues aimed to ease the prediction of the result of behavioral synthesis: interpretation of VHDL constructs, VHDL execution models and scheduling of VHDL descriptions.

Most behavioral synthesis tools are sensitive to the VHDL writing style. In many cases, the synthesis results may be influenced by changing the writing style of the behavioral description. Section 1 details the interpretation of the main VHDL constructs that may influence the synthesis results.

Each behavioral synthesis tool performs specific high level transformations that may change the initial behavior, for example the order of I/O signals. Section 2 gives the main interpretations of VHDL behavioral models.

Scheduling is the main added value of behavioral synthesis. However, many scheduling algorithms exist. Section 3 presents the main scheduling algorithms used by behavioral synthesis tools.

3.1 INTERPRETATION OF VHDL DESCRIPTIONS

This section deals with VHDL behavioral descriptions for synthesis. Most VHDL-based behavioral synthesis tools start from a VHDL subset restricted with a well defined writing style. The main concepts accepted by these behavioral synthesis tools will be outlined. For each concept we will overview the main possible interpretations for behavioral synthesis. We note that this study is general and will not be restricted to a specific synthesis tool.

3.1.1 Entity/architecture/process

A VHDL model consists basically of entities and architectures. An entity defines the interface between a design and its environment through a list of input and output ports. The entity declaration generally defines the external view of the design in question. It is still a *black box* with ports that may be specified, but there is no information about how it maps inputs to outputs. In the example of figure 3.1. the entity bubble has an output port called *data_out* and three input ports: *data_ready, data_in* and *start*. The architecture describes the relationships between the inputs and outputs of the design. In other words, it defines how the entity behaves or what it is composed of. One can model the design with any combination of the description levels : algorithmic, RTL, logic or structural. An architecture consists of declarations and concurrent statements. In the example of figure 3.1, the architecture called *behavior* of entity *bubble* contains only one concurrent statement represented by the process *Sort*. A process is in effect an infinite loop between the begin and end process statements.

The compilation unit for most VHDL behavioral synthesis systems is a VHDL entity/architecture pair that contains a single process. However, the description of a complex, real-time system is more complicated. Besides, blocks that may be compiled with behavioral synthesis tools, a system specification includes several components needed to interface the system to its environment. For example, a telephone answering machine needs to be able to communicate with the telephone exchange as well as internal components such as the tape deck and so on.

A complex system specification should firstly be partioned into a set of interconnected processes, as shown in figure 3.2. There are two kinds of processes: interfaces and computation processes. Interfaces are generally designed to link

VHDL modeling for behavioral synthesis   69

```
entity bubble is                                    For i in 1 to 256 Loop
    port (data_ready : in Boolean;                      Wait until data_ready;
        data_in : in Integer;                           data_out <= ram(i);
        data_out : out Integer;                     end Loop;
        start : in Bit;                             end write_mem;
        - - other ports );                      Begin
end bubble;                                         - Algorithmic part (one process)
                                                Sort: Process
architecture behavior of bubble is                  Variable ram : memory;
    - declarative part of the architecture          Variable tmp,t1,t2: Integer;
    Type memory is Array (1 to 256) of Integer;     Begin
                                                        Wait until (start = '1');
    Procedure read_mem(                                 read_mem(data_ready,data_in,ram);
        data_ready : in Boolean;                        For i in 1 to 256 Loop
        data_in : in Integer;                               t1 := 257;
        ram : out memory) is                            While t1 > i Loop
Begin                                                       t2 := t1 - 2;
    For i in 1 to 256 Loop                                  t1 := t1 - 1;
        Wait until data_ready;                              If ram(t1) < ram(t2) Then
        ram(i) := data_in ;                                     tmp := ram(t1);
    end Loop;                                                   ram(t1) := ram(t2) ;
end read_mem;                                                   ram(t2) := tmp ;
                                                            end if;
                                                        end Loop;
    Procedure write_mem(                                end Loop;
        data_ready : in Boolean;                        write_mem(data_ready,data_out,ram);
        data_out : out Integer;                     end Process Sort;
        ram : in memory) is                     end behavior;
Begin
```

Figure 3.1   Excerpt VHDL model of a Bubble sort

the system to the external world through fixed communication protocols (clock rate, physical characteristics of the signal, etc.). The specification of such processes is generally given at lower levels (RT, gate or physical level). This corresponds to a VHDL description that contains clocks and other synchronization signals. The computation processes represent components that are synchronized through events and described at the behavioral level. The specification provides only the operations performed and their order. In VHDL, they may

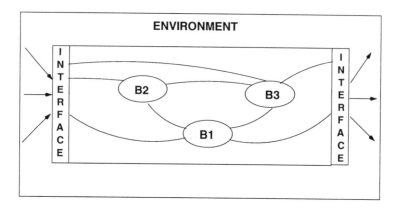

Figure 3.2 Complex system specification example

be described by processes containing unstructured wait statements and nested loops.

3.1.2 Behavioral description

The behavioral description is the part of the VHDL model accepted by a behavioral synthesis tool during a synthesis session. This is generally restricted to a single VHDL process describing the behavior as an algorithm. A VHDL process consists of a declarative part which can contain mainly local variables and a process body including sequential statements. These statements are similar to those used in procedural programming language such as Ada, C or Pascal. During behavioral synthesis, this description will be mapped into an architecture composed of a datapath and a controller. This is generally encapsulated in a VHDL entity/architecture pair where:

- The ports correspond to the signals used by the process to communicate with its environment.

- The architecture is made of two parts; a datapath and a controller.

- The controller is generally represented as an FSM executing all the commands of the initial behavior.

- The datapath is a netlist of modules able to execute the transfers of the behavioral description.

VHDL modeling for behavioral synthesis   71

Although, the correspondence between the behavioral model and the architecture is not always obvious, the writing style of the behavioral description has generally a large influence on the resulting architecture. This will be discussed in the rest of this chapter.

3.1.3 Conditional statements

VHDL supports the usual conditional statements to represent conditional behavior, such as if and case. The If statement is of the form:

```
if boolean expression then
   – statements
elseif boolean expression then
   – statements
....
else
   – statements
end if;
```

The case statement is of the form:

```
case identifier
   when case1 => –statements
   when case2 => –statements
   ...
   when others => –statements
end case;
```

The way these conditional statements are interpreted by behavioral synthesis may have a large influence on the synthesized architecture for both the controller and the datapath. This impact will be illustrated using the following example:

72 CHAPTER 3

```
if (X<Y)
    then Y:= Y - X ;
    else X:= X - Y ;
end if;
```

Figure 3.3(a) and figure 3.3(b) show two possible implementations. In the first one, the condition is evaluated in the controller using specific resources. In the second one, the condition is executed in the datapath. On the controller side, a conditional construct will produce:

- Extra transitions in the corresponding FSM.

- Extra input signals if the condition includes objects used in the datapath. This corresponds to the status lines X and Y in figure 3.3(a).

- Arithmetic and logical operators if the condition is complex.

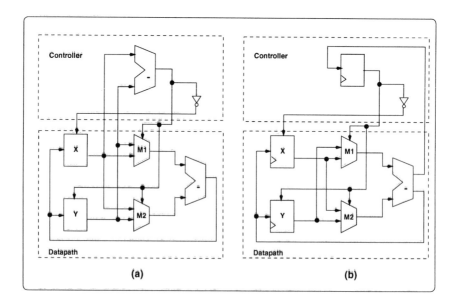

Figure 3.3 Interpretation of conditional constructs

On the datapath side, a conditional construct result in:

VHDL modeling for behavioral synthesis

- Extra connections (MUXs) when a destination may have several inputs, M1 and M2 in figures 3.3(a) and 3.3(b).

- Functional units in the case where the synthesis tool decides to put part or all the execution of the condition in the data path. For the example of figure 3.3(b), the condition $X < Y$ may be executed on the same FU as the operation $-$. However, this will result in an extra execution cycle.

Conditional constructs result in several branches that are mutually exclusive. During one execution path of the design, only one branch gets executed based on the evaluation condition. The user must be careful when using these constructs since the writing style can influence the performance of the synthesized design. This is illustrated in the VHDL descriptions of figure 3.4.

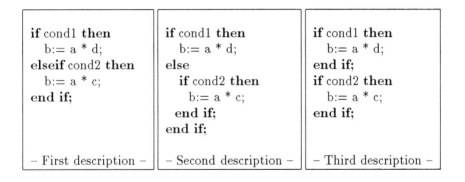

Figure 3.4 Writing styles for a conditional construct

Suppose now that cond1 and cond2 are mutually exclusive and never occur together. In this case, the three descriptions behave the same, but the three descriptions will result in three different implementations. The first description will result in the most optimized design in terms of execution time (one control step), in terms of resources (one multiplier) and two mutually exclusive execution paths triggered by $\{cond1, cond2\}$. In the second description, the multiplier can also be shared. However, the execution time will depend on the scheduling algorithm. It may vary from one to two control steps. In fact, some scheduling algorithms schedule the bodies of conditional constructs in different control steps. The second description produces three transitions in the controller FSM triggered by three conditions$\{cond1, \neg cond1 \wedge cond2, \neg cond1 \wedge \neg cond2\}$. The third description consists of four transitions triggered by $\{cond1 \wedge cond2, cond1 \wedge \neg cond2, \neg cond1 \wedge cond2, \neg cond1 \wedge \neg cond2\}$.

Only few behavioral synthesis tools [CBH+91, JCG94] will detect that the first condition $\{cond1 \wedge cond2\}$ is always false. If this transition is not eliminated, the description will necessitate two multipliers. In the case where the scheduling algorithm is constrained by one multiplier, two control steps are usually required to execute the conditional statement.

3.1.4 Synchronization statement

The wait statement is the key to effect the synchronization of concurrent processes. The wait statement has the form:

> Wait [sensitivity_clause] [condition_clause] [timeout_clause]

The sensitivity clause defines the set of signals to which the wait statement is sensitive. That is, execution continues if there is an event on any signal in the sensitivity list. For example the statement:

> *Wait on S1 until S2;*

means that execution continues if $S1$ changes and the new value of $S2$ must be true. This clause is not supported by most behavioral synthesis tools.

The condition clause specifies a condition that must be evaluated as true to continue execution. This clause can be used to ensure synchronization between processes explicitly. Take, for example the description of the two processes shown in figure 3.5. Process $P2$ has to wait for process $P1$ assigning *data_ready* to continue execution. This clause is supported by all behavioral synthesis tools, but several tools restrict this clause to be a simple *clk'event*; e.g:

> *wait until clk'event and clk ='1';*

From a behavioral synthesis point of view, a wait statement may be a synchronous or an asynchronous statement. A synchronous statement is evaluated on clock edges only. A typical synchronous wait statement is :

> *wait until condition and rising_edge(clk);*

VHDL modeling for behavioral synthesis

An asynchronous wait statement may be evaluated independently from the clock. This kind of statement is difficult to handle in a clean way in behavioral synthesis. In fact, in the RTL model produced by behavioral synthesis, all the wait statements are synchronous. This makes difficult the correspondence between a behavioral description and the resulting RTL model.

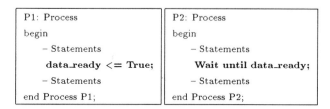

Figure 3.5 Process synchronization

The timeout clause causes the process to be suspended until a timeout interval has expired. The timeout interval is given in absolute time. Some behavioral synthesis systems [ROJ94, NBD92] ignore this clause because the synthesis tool cannot guarantee that the implemented circuit will exhibit an arbitrarily specified delay characteristic. To overcome this problem, the IBM VHDL synthesis system [Sau87] associates an attribute to specify the cycle time and each time expression is converted to control steps. For example the wait statement *Wait for 2*cycle_time;* is converted to two control steps during scheduling.

In VHDL, a wait statement specifies a control state of the behavioral description. During the simulation of a given process, when a wait statement is reached, the execution of the process is stopped until the condition to resume the wait is fulfilled. In behavioral synthesis, the wait statement is generally interpreted as a sequence break. It introduces a new state in the controller. A wait statement is an implicit state imposed by the designer through the behavioral description. The code executed between two successive wait statements is called *thread*. A behavioral VHDL description is equivalent to an FSM where the states correspond to the wait statements and the transitions to the different execution paths (or threads) between two wait statements (code executed during a simulation step). In fact, the only observable points in the behavioral description are wait statements in which signals are evaluated. The code between two wait statements may include complex control statements including loops. It will be decomposed into several control steps during scheduling. The different scheduling modes of VHDL descriptions will be detailed in section 3.2.

3.1.5 Iterative statements

VHDL supports repetitive operations by using loops, similar to standard programming languages. The loop types include unconditional loops, *While* and *For* loops. Related statements used with loops are the *Next* and *Exit* statements.

The Next statement causes a jump to the beginning of a loop. The *Exit* statement allows multiple exit points in a Loop.

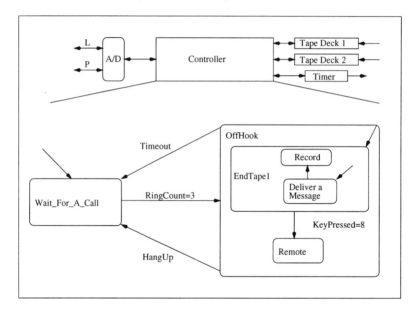

Figure 3.6 Telephone answering machine; (a) block diagram of system; (b) StateChartlike model of controller.

Loops allow to express complex behavior. This can be shown by the example of a telephone answering machine. Figure 3.6(a) shows a block diagram of such a machine. The system contains a controller, two tape decks, a timer and an interface with the telephone network via an A/D converter. Figure 3.6(b) shows the state hierarchy of the controller part of the telephone answering machine. The entry state of the system is the state *Wait_For_A_Call*. In this state, the system is initialized and the flow of control waits until three rings have been received from the telephone exchange. When this event occurs, a transition to the state *OffHook* is made. In this state, a message is delivered and the

VHDL modeling for behavioral synthesis

caller can either leave a message, remotely play back all previously recorded messages, or hang up. If the caller hangs up at any stage during the process, an immediate transition to the state *Wait_For_A_Call* must be made. A similar transition is made if any of the timeout restrictions is not adhered to.

(a)	(b)

(Loop behavior
  (Loop Wait_For_A_Call)
  (Loop Get_3_Rings)
)
(Loop OffHook
  (Loop Remote
    (Loop Get_3_Digits)
    (Loop ManualControl)
)
)
)
	E1 : Set of expressions
	1 OffHook : Loop
	2 Get_3_Digits : Loop
	3 Wait Until (KeyPressed=0);
	4 TimeOut:= FU_REM(ElapsedTime,20);
	5 Wait Until ((ElapsedTime=Timeout) Or
	        (KeyPressed/=0) Or (HAngUp='1');
	6 If ((HangUp='1') Or (ElapsedTime=Timeout))
	7 Then Exit OffHook; End If;
	8 NextDigit:= PasswdROM(DigitCount);
	9 If (KeyPressed/=NextDigit)
	10 Then FalsePasswd:=1; End If;
	11 DigitCount := DigitCount+1;
	12 If (DigitCount=3)
	13 Then Exit Get_3_Digits;
	    End If;
	  End Loop Get_3_Digits;
	  E2 : Set Of expressions;
	End Loop OffHook
	E3 : Set of expressions;

Figure 3.7 VHDL representation of the controller; (a) Loop hierarchy; (b) Code segment.

The controller can be modeled as a single VHDL process containing seven nested loops organized as shown in figure 3.7(a). The full VHDL description of this answering machine contains 180 lines. For clarity, we concentrate on a single loop, namely the *Get_3_Digits* loop (figure 3.7(b)). To represent the rest of the code, we have added a set of dummy nodes before and after the loop (statements E1, E2 and E3), as well as showing the loop *OffHook*. During the execution of any of these loops, if there is a global exception (the caller hangs up, for example), the loop hierarchy must exit and the control must be passed to the *Wait_For_A_Call* loop. After each wait statement, all global exceptions must be tested (see operation 6 of figure 3.7(b)).

After showing the importance of iterative constructs for complex design modeling in VHDL, we will concentrate on the way they are treated in the synthesis process.

Generally, loops need the application of sophisticated algorithms by behavioral synthesis, since they usually dominate the execution time of the behavioral description. In order to improve the synthesized design, complex transformations are generally needed. These transformations may be applied during VHDL compilation or during the scheduling phase. These transformations will be illustrated by an example. Consider a loop where each iteration takes S control steps to execute. There are several schemes for executing it. The more classical way, is to execute it sequentially. Each S steps, a new iteration begins. Suppose that the average number of iterations is N, then the total execution time of the entire loop is $S \times N$. Figure 3.8(a) shows the sequential execution of a finite loop consisting of 9 iterations. The advantages of such scheme is that it produces a simple controller and datapath. However, the execution time of the loop can be reduced, since loops can exhibit potential parallelism beyond the iteration boundaries which can be exploited by executing several iterations of the loop concurrently. In a sequential scheme, we look to only one iteration at a time and this parallelism cannot be detected.

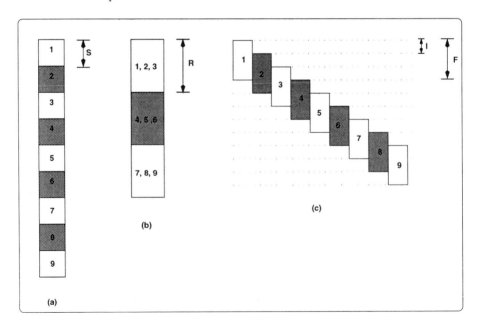

Figure 3.8 Loop execution, (a) sequential, (b) unrolling, (c) folding

The second scheme consists of *unrolling* the loop partially or totally, this results in a loop with a larger body but fewer iterations. This has the benefits of a greater flexibility for optimizing the execution time of the transformed loop body. For example figure 3.8(b) shows that every three iterations are unrolled into one *super-iteration*. An efficient schedule of a super iteration produces R control steps where $R < S \times 3$. The total execution time of the loop becomes $R \times 3$ which is less than $S \times 9$ control steps.

The last scheme exploits intra-loop parallelism by introducing partial overlaps between the execution times of successive loop iterations. We call this *loop folding*. In this scheme, successive iterations are initiated every I control steps. I is called *initiation interval* and is defined as the difference between the start times of two successive iterations. Figure 3.8(c) shows a loop with a body that takes F control steps, F is called *latency*. In this scheme, the total execution time of the loop is $F + (N - 1) \times I$ control steps.

After introducing the different schemes for loop executions, we will discuss their suitability for the different kinds of iterative constructs.

Infinite loops

They are of type:

```
infinite:loop
    - - loop body
end loop infinite;
```

A typical example of an infinite loop is a filter for digital signal processing applications which perform the same set of operations on every sample of input data stream. The loop execution time of such behavior dominates the total execution time of the design, so, optimizing the throughput of the loop body is essential to the performance of the design. In this case, loop unrolling is not applicable since the loop count is infinite. However, loop folding is preferable when the design throughput is critical.

Loops with known bounds

They are of type:

```
for i in 1 to N loop
   - - loop body
end loop;
```

where N is an integer known at compile time. This kind of loops is the most studied one for parallelization purposes. All the schemes mentioned above are applicable.

Loops with one unknown bound

They are of type:

```
for i in 1 to N loop
   - - loop body
end loop;
```

where N in unknown at compile time. For the same reason as for infinite loops, loop unrolling is not applicable.

Data dependent loops

They are of type:

```
while condition loop
   - - loop body
end loop;
```

where *condition* is an expression depending on external signals. In this case neither loop unrolling nor loop folding is applicable. However, several techniques exist for pipelining such loops. Among them *Rotation scheduling* [CLS93], a transformation-based approach for loop pipelining. It repeatedly up or down rotates a schedule to get a more compact one under resource constraints. This technique is applicable only in data flow oriented designs. *Pipeline Path-based scheduling* [RJ95b] is another technique for pipelining data dependent loops. It considers that a data dependent loop is executed 0, 1 or 2 or more times. So, it generates execution paths assuming the loop executes 0 times, once and twice. This technique is not unlike that of loop unrolling [GVM89]. Unrolling loops twice allows us to detect any inter-dependencies that may exist between different iterations of the same loop. This technique will be detailed in section 3.3.2. One should note that the optimization of a loop throughput increases the control costs.

The combination of loops, conditional statements and waits is restricted in several behavioral synthesis tools. These restrictions are generally imposed by the intermediate representation and the scheduling algorithm used.

3.1.6 Procedures and functions

VHDL functions and procedures are VHDL subprograms used to simplify the coding of repetitive or commonly used portions of code. If a certain block of code is used repeatedly in a design to implement a particular function, it can instead be placed in a function or procedure once and then referred to by name as needed. The main difference between these two types of subprograms is that procedures can have multiple outputs, while functions should have only one output. Variables used in a function or procedure are only valid within that function or procedure. within a process, we can define a hierarchy of procedure calls. This corresponds to the well known procedure call mechanism in modern programming languages. Procedure and function calls may be handled in three different ways by behavioral synthesis tools. These schemes will be illustrated using the example shown in figure 3.9:

- The first scheme consists of the in-line expansion of the call. The call is replaced by the code of the corresponding procedure or function. Of course flattening the procedure increases the size of both the behavioral description and the generated architecture (mainly the controller). However, the operations inside the procedure or function may share resources with the operations outside. Figure 3.10(a) shows the transformed VHDL description after in-line expansion of procedure P. In this case, the datapath may contain only one multiplier. However, the controller becomes more complex.

```
P1: Process
    ___
    Procedure P(a: in integer;b: out integer)
    begin
        c:= a * d;
        b:= a * c;
    end P;
begin
    wait until rising_edge(clock);
    P(x,y);
    z:= y*k;
    wait until rising_edge(clock) ;
    P(z,y);
end process P1;
```

Figure 3.9 VHDL description example

- The called procedure or function is compiled as an extra unit (independent processor) as shown in figure 3.11. In this case the call is replaced by a communication protocol to transfer the parameters between the main module (VHDL process) and the processor corresponding to the called procedure or function.

- The call is interpreted as an operation corresponding to an existing functional unit in the datapath as shown in figure 3.12. This model allow the reuse of complex components called *coprocessors*. In this case some restrictions are often imposed such as a fixed execution model (the body of the procedure or function can be scheduled in a fixed number of control steps).

3.1.7 Signals and Variables

A process communicates with its environment using signals. VHDL signals correspond to actual design signals and are structural elements that can be used inside processes [CST91]. When a signal is assigned multiple values, a resolution mechanism defines the actual value of a signal.

after clauses in signal assignment are generally prohibited or ignored in most behavioral synthesis systems. Signal assignments have a delayed effect in VHDL.

VHDL modeling for behavioral synthesis

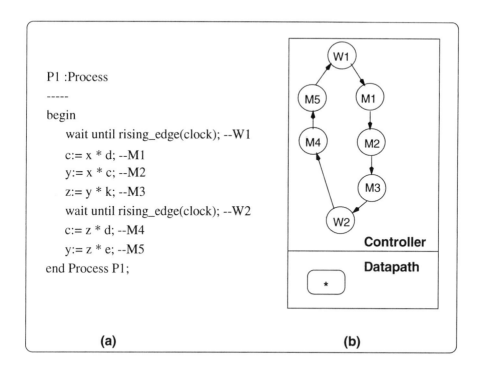

Figure 3.10 In-line expansion of the procedure

After an assignment, a signal will take its new value only after a δ cycle; i.e during the next simulation cycle or when a wait statement is encountered during simulation.

In VHDL, when a signal is assigned twice within an execution thread (an execution path between 2 wait statements), the first assignment will be ignored. These assignment should be detected and removed during synthesis. This kind of assignment is generally useful for assignment of default values. For the example in figure 3.13, x is completely defined in both descriptions. However, it is easy to see that the use of redundant signal assignments may induce a drastic reduction in the size of the description.

In VHDL signals are the primary data items. Signals have a time history of values. There are two types of signals: bus and register. The main difference is that register signals have memory. They retain the value they were last assigned by a driver.

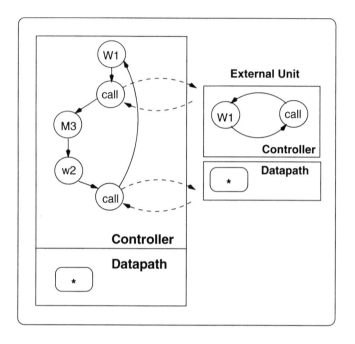

Figure 3.11 Procedure as an independent module

All the signals used in a behavioral synthesis (main synthesized process) will be mapped to a port or to a storage element during behavioral synthesis.

Even the synthesized process is mixed with a nonsynthesizable code within a VHDL architecture, the entity corresponding to the resulting architecture should be synthesized automatically. A global analysis similar to the one performed by RTL synthesis may be executed in order to decide whether the signal should be mapped to a storage unit or not.

As in any programming language, variables in VHDL retain their values until a new assignment is made.

In VHDL, variables have the same behavior as those in programming languages. Properly used they can increase the speed and efficiency of high-level simulations.

Variables have limited scope. Processes live forever, so variables defined in a process do not lose their value. However, a variable defined in a procedure or

VHDL modeling for behavioral synthesis

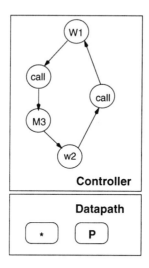

Figure 3.12 Procedure as a coprocessor

function will not be remembered between different calls to that procedure. The synthesis of variables depends heavily on the synthesis tool used. Conceptually, VHDL variables have not to be stored. These are only used for intermediate computations.

Variables and signals may be used as values or as storage elements. For signals, this classification is similar to the one used in logic synthesis. When the value of a signal should be stored more than one cycle for a given process, the signal should be mapped to a storage element. Otherwise, the signal is considered as a wire. In the rest of this section, we are concerned only with signals that have to be mapped into storage units.

The synthesis of variables and signals consists of mapping them onto registers. Variables may also be mapped onto register files or memories depending on the synthesis tools. There are many ways to perform this task:

- Each variable is mapped onto a single register. This has the advantage of preserving the same variable name before and after synthesis.

- Several variables can share the same register. This task is performed by an evaluation of the lifetime of variables, defined as the time interval between the first assignment of a variable and its last use. Variables with

```
wait until cond1;
case y
    when 0 =>
        x0<=b0;
        x1<=a1;
        x2<=a2;
        ......
        xn<=an;
    when 1 =>
        x0<=a0;
        x1<=b1;
        x2<=a2;
        ......
        xn<=an;
    when 2 =>
        x0<=a0;
        x1<=a1;
        x2<=b2;
        ......
        xn<=an;
    ............
    when n =>
        x0<=a0;
        x1<=a1;
        x2<=a2;
        ......
        xn<=bn;
    when others =>
        x0<=a0;
        x1<=a1;
        x2<=a2;
        ......
        xn<=an;
end case;
wait until cond2;
        (a)
```

```
wait until cond1;
x0<=a0;
x1<=a1;
x2<=a2;
......
xn<=an;
case y
    when 0 => x0<=b0;
    when 1 => x1<=b1;
    when 2 => x2<=b2;
    ............
    when n => xn<=bn;
    when others =>
end case;
wait until cond2;
        (a)
```

Figure 3.13 (a) Original signal assignments, (b) Redundant signal assignment

non-overlapping lifetimes are then mapped onto the same register in such a way as to reduce the total number of registers. There are severals techniques performing this task, we can cite the left-edge algorithm [KP87] which binds variables sequentially to one register at a time. This algorithm starts by computing variable lifetimes, then it sorts the list of variables in ascending range with the start time of their lifetime interval and in descending range with their last time of their lifetime interval. An example of mapping variables to registers using this algorithm is shown in figure 3.14. Other algorithms performing this task can be found in [Sto91, GDWL92].

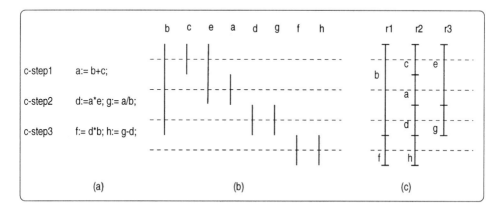

Figure 3.14 Mapping variables to registers using the left-edge algorithm, (a) description example, (b) sorted variable lifetime intervals, (c) resulting registers

- Variables are mapped to memories. This task is performed by firstly mapping these variables to register and then grouping the registers into memories with single or multiple access port. Registers can be grouped into the same memory as long as the number of register accessed in the same control step does not exceed the number of ports. Several authors have considered this problem [AC91, KL93]. The problem of mapping arrays to memories will be addressed in the next section.

3.1.8 Data typing

VHDL is a strongly typed language. VHDL supports signals, variables and constants, all of which may have any number of predefined or user-defined data types. VHDL like other modern languages comes with predefined data types, such as integers and permits users to define their own data structures. Types in VHDL include *scalar types, composite types, access types* and *file types*.

Scalar types

Scalar types have single values and include integers, floating-point and enumerated types. These types do not affect behavioral synthesis.

Composite types

Composite types include records and arrays. Records are data structure defined by the user where their elements can be of different types. Generally, the compiler decomposes records into their components.

Arrays are data structures where each element of the array is of the same type. Signal arrays do not allow dynamic indexing during assignment. That is, an index must always be known at compile time. Variable arrays are very useful and very important for hardware descriptions. In contrast to arrays of signals, during assignment their indexes can be variable or fixed. Generally, arrays of variables are mapped onto multi-port memories.

The specification of an array is analogous to that of a memory as the array is composed of a number of rows of data. For example, array *ram* in the example of figure 3.1 can result in a 16-bit-wide 256-word memory (in the case of attribution of a size of 16 bits to an integer type). Many works [RGC94, ST95] treat the problem of mapping arrays in a behavioral description onto memories. These approaches try to associate groups of arrays to the same memory or to flat the same array onto different memories. These associations are done while minimizing area and maximizing performance. An example of array mapping is shown in figure 3.15. Array A is split into two memories M1 and M2. Array B corresponds to a memory M3 of the same size. Arrays C and D are grouped together in the same memory M4.

Access and File types

Many languages permit the definition of variables that refer to other variables. These variables contain the address of the variables being referenced. In VHDL these variables are called *access types*. In Programming languages such as Pascal or C, these variables are called pointers. Access types are used only for behavioral simulation. There is no hardware equivalent to them since they need dynamic memory allocation. Almost all synthesis systems do not support this type. File types and file objects are not supported by almost all synthesis

VHDL modeling for behavioral synthesis 89

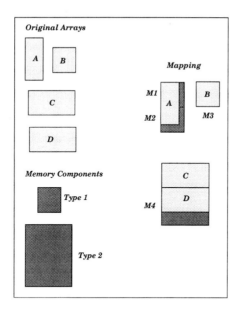

Figure 3.15 Array to Memory Mapping

systems. The reason is that a file object may have associated with it an external file in the host system.

This object however can be helpful to model complex memory contents for example a ROM which contains large data. Assigning this ROM manually as an array constant in the behavioral description is a waste of time and prune to error. An interesting way to solve this problem is to insert a pre-processing step called elaboration [OM96] which transforms a nonsynthesizable description using file objects to a synthesizable description in which the file objects are removed.

3.2 BEHAVIORAL VHDL EXECUTION MODES

During the design process, the designer may need to predict and/or to understand the results of behavioral synthesis. This can help the designer to debug, simulate and may be reuse the resulting architecture. As stated above, this is not an obvious task because behavioral synthesis makes use of complex transformations that may change the I/O timing of the behavioral description. In addition, different behavioral synthesis tools make use of different VHDL subsets, different writing styles and different interpretations of VHDL constructs (e.g procedures, conditional statements, loops, ...).

This section introduces the main execution modes of behavioral VHDL descriptions. Six modes will be defined and classified according to the scheduling of VHDL execution threads and the scheduling of I/O operations.

3.2.1 Scheduling of VHDL execution threads

An execution thread is defined as the code executed between 2 successive wait statements during simulation. As stated in section 3.1.4, a behavioral VHDL description is equivalent to a high-level FSM where the states correspond to the wait statements and the transitions to the different execution paths or threads. This is illustrated by figure 3.16.

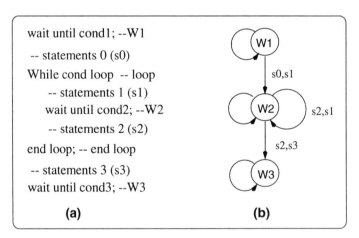

Figure 3.16 High-level FSM representation of a behavioral VHDL description

Figure 3.16(b) shows the high-level FSM representation corresponding to the behavioral VHDL description of figure 3.16(a). The FSM is composed of three states (W1,W2,W3) and six transitions. Each transition corresponds to a thread. A thread corresponds to a computation step (see section 1.2). It may

VHDL modeling for behavioral synthesis 91

include complex computation and data dependent loops. One of the central functions of behavioral synthesis is to :

- Extract the threads from the behavioral description
- Schedule the threads, i.e splitting them into a set of clock cycles.

These two sub-tasks constitute what is called scheduling in behavioral synthesis. Scheduling methods will be detailed in section 3.3. This section is restricted to the presentation of the different modes of thread scheduling. There are mainly three classes of thread scheduling. These are defined according to the results produced by scheduling which may be composed of a single control step, a loop free sequence of control steps or a set of sequenced control steps that may include loops.

The first class is called cycle-fixed scheduling mode. Each thread is scheduled in a unique control step that can be executed in one single clock period. In this case, the interpretation of the VHDL description is very closed to the one performed by RTL and logic synthesis tools. In this mode and in order to be synthesized, a VHDL description should follow a specific writing style that imposes the following restrictions on threads:

- Threads cannot include data dependent or infinite loops.
- Threads cannot include multi-cycle operations.
- The clock period should be long enough to allow the execution of all the operations of the most complex thread.

The main advantage of this mode is the closeness between the behavioral description and the produced RTL model. The writing style makes it very easy to have the same behavior when simulating the two VHDL models. This way, a single testbench can be used for the simulation of both descriptions. The differences between the two models are related to synchronization. These correspond to the automatic introduction of Reset signals [Kna96] and to the synthesis of asynchronous waits. If the asynchronous waits are avoided and the insertion of resets is taken into account, a unique testbench may be used for both simulations.

The second thread scheduling class is called super state scheduling mode [Kna96]. Each thread may be mapped onto a sequence of control steps. The number of control step is fixed during the scheduling process. This style is a large extension to RTL synthesis: complex threads may be decomposed in order to be executed during several clock periods. The only restriction imposed by this scheduling mode on the VHDL description is to forbid data dependent and infinite loops within threads.

The main advantage of this scheduling mode is the possibility to explore several solutions starting from the same behavioral description. This model changes the latency of the design by splitting behavioral computation steps into several control steps. This makes harder the correspondence between the behavioral description and the produced RTL one. Besides the differences listed for the cycle fixed mode, the latency should be taken into account in order to be able to use a single testbench for the simulation of both behavioral and RTL models. This scheduling mode handles behavioral descriptions written for the cycle fixed mode.

The third thread scheduling mode is called behavioral state scheduling mode. Each thread is scheduled into an FSM where each transition may execute in a single clock cycle. In this case, the resulting schedule may include loops and conditional transitions. This is the most general scheduling mode. It imposes no restrictions on the writing style of the behavioral description. This scheduling mode offers more freedom when writing behavioral descriptions and allows more architectural exploration starting from the same behavior. However it is harder to implement. In addition the correspondence between the behavioral and the RTL description becomes even harder. In order to be able to simulate both models using the same test bench, the communication between the design and the external world should be nonsensitive to the scheduling. A typical solution for solving this problem is to use handshakes. Chapter 7 gives an example of using this mode within a methodology allowing the use of a unique testbench for both behavioral and RTL descriptions. This mode handles behavioral description written for the previous modes.

3.2.2 Scheduling of I/O operations

Most scheduling algorithms perform some local and/or global reordering of the operations of the behavioral description. This transformation consists of moving operations within one thread and/or from one thread to another without changing the behavior of the description. Such operation re-ordering may be useful for resource sharing. The application of these transformations to I/O operations may change the external behavior of the design under synthesis.

There are two strategies when scheduling I/O operations. The first is called fixed I/O mode and preserves the order of I/O operations. The second allows re-ordering of I/O operations and is called free-floating I/O mode [Kna96].

In cycle fixed I/O mode, several strategies exist for scheduling I/O operations within one thread. In [Kna96], when scheduling a thread in super state mode, inputs are scheduled in the first cycle and outputs in the last cycle. This strategy imposes extra restrictions on the writing style (e.g position of output within a loop).

In free-floating mode, I/O operations are handled like other VHDL statements. They are reordered in order to produce a schedule that meets time and area constraints. The advantage of this mode is that it allows for a larger design space exploration starting from the same behavioral description. But, such a mode makes harder the correspondence between the behavioral description and the corresponding RTL one.

3.2.3 VHDL interpretation modes

The combination of thread and I/O scheduling modes produces six synthesis styles summarized in table 3.1.

I/O Scheduling Mode	Thread Scheduling Mode		
	Cycle Mode	*Super State Mode*	*Behavioral State Mode*
Fixed I/O Mode	1. Cycle Fixed I/O	2. Super State Fixed I/O	3. Behavioral State Fixed I/O
Floating I/O Mode	4. Cycle Floating I/O	5. Super State Floating I/O	6. Behavioral State Floating I/O

Table 3.1 Interpretation modes of behavioral VHDL

Each of these styles corresponds to a specific interpretation mode of the VHDL description. As explained above, each of these modes may have advantages and disadvantages with regards to the freedom of the writing style and the ease of the correspondence between the behavioral model and the synthesized RTL description.

3.3 SCHEDULING VHDL DESCRIPTIONS

The first step in high level synthesis is to derive an internal representation from the input VHDL description. There are several types of representations [KD92, CT89, Wol91, Sto91, Jon93, GW92]. The choice of the representation depends on the synthesis algorithms to be used and particularly the scheduling one. Scheduling is the most complex task in behavioral synthesis and involves fixing the execution order of operations in a behavioral description. Scheduling algorithms can be classified into two major classes [RJ96a]: Data Flow-Based Scheduling (DFBS) and Control Flow-Based Scheduling (CFBS). We define:

Data Flow-Based Scheduling, Control Flow-Based Scheduling as all the scheduling algorithms using a data flow graph or control flow graph respectively in its strict sense as the underlying internal representation for scheduling.

To show the main differences between the two classes, let us take two scheduling approaches one using a data flow graph (DFG), the other a control flow graph (CFG) as shown in figure 3.17. The scheduling of a DFG consists of assigning

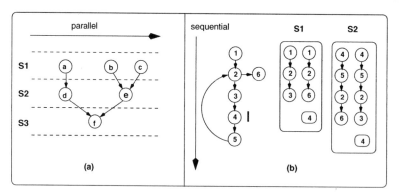

Figure 3.17 (a) Data-flow scheduling. (b) Control-flow scheduling.

each *operation* to a control step in such a way that all given constraints are met. The constraints fix either the number of control steps or the amount of resources to be used. An operation can be assigned to only one control step. A DFG is the most parallel representation. Parallelism is derived from the data dependencies among operations. DFBS sometimes have to reduce the amount of parallelism due to the given constraints. In contrast to DFBS, CFBS use a CFG as the underlying representation for scheduling. A CFG is by nature a

VHDL modeling for behavioral synthesis

sequential representation of a behavioral description. So, CFBS have to extract the maximum of parallelism and at the same time meet the given constraints. CFBS consist of assigning a *sequence* of operations to a control step, then an operation can be assigned to more than one control step as operations 3 and 6 in figure 3.17.

A DFBS is equivalent to a two dimensional optimization problem. So, we say that a DFBS is more efficient than another one if :

- with the same number of resources, it uses less control steps to schedule a given description or
- with the same number of control steps, it uses less resources.

When a behavioral description contains control constructs, this assumption is no more valid. The major scheduling techniques assume that all the branches of a control construct have the same probability to be executed and this is not always the case. We will see in section 3.3.2, by introducing a new metric [BDB94] which consists of an estimation of the execution time of an algorithm, that minimizing the number of control steps may not be equivalent to minimizing the execution time of an algorithm.

DFBS are defined to operate just on single basic blocks, one section of straight-line code with only one entry and one exit point. Since VHDL support conditional constructs such as *if-then-else, case* and loops, some DFBS consider those constructs during the scheduling step. They partition the graph into basic blocks joined by extra nodes such as *fork* for *if* and *case,* and *join* nodes for *end if* and *end case*. For control flow intensive specifications, using DFBS techniques decreases the performances of the scheduled design in two ways:

- This partitioning creates extra control steps between basic blocks.

- Two branches of a conditional construct are scheduled in the same number of control steps. For example, suppose there is an if statement within a loop:

> **while** (condition)**loop**
>   **if** BoolExpr **then**
>   block of statements
>   **else**

>   single statement
>   **end if;**
>   **end loop;**

Assuming that the execution of the block of statements takes 10 control steps, it is easy to see that when BoolExpr is false, each loop cycle takes only one control step while if BoolExpr is true, it takes 10 control steps. DFBS approaches schedule the critical path first and afterwards add all the other operations in parallel with this critical path. Consequently, all paths will unnecessarily take the same number of control steps.

- For repetition constructs, DFBS approaches try to schedule the innermost loop body and the corresponding test expression first, and then to proceed outwards treating the loop as a single node.

3.3.1 Data flow-based scheduling

As defined in chapter 1, a data flow graph represents a partial order $\prec$ on all operations $v \in V$, specifying precedence constraints. So, $v_i \prec v_j$ means that operations represented by v_i have to be executed before v_j can start.

Based on the definition in [Sto91], a *schedule* of the data flow graph is a mapping

where:
$$\sigma : V \longrightarrow S$$
$$v_i \longmapsto \sigma(v_i)$$
$$v_i \prec v_j \Rightarrow \sigma(v_i) < \sigma(v_j).$$

S is the set of control steps. Most of the scheduling algorithms are optimized for data flows. The main difference between all these algorithms is the choice whether an operation is scheduled in a given control step, from a set of control steps and/or a set of functional units. Data flow scheduling is a two dimensional problem: time and area. Iterative approaches impose constraints on one dimension and try to optimize the other. However, global techniques such as ILP [GE92] try to optimize both dimensions simultaneously. So, for iterative approaches, there are two types of scheduling problem [WC95], the first fixes the number of control steps and tries to minimize the number of functional units needed to execute the operations within the given time. We call this a time-constrained scheduling. The second assumes a number of modules and

VHDL modeling for behavioral synthesis

tries to minimize the number of control steps. This approach is called resource-constrained scheduling. Consider the signal assignments shown in figure 3.18 which might be part of a VHDL process description. This description will be

```
x <= (a + b) * (c + d) * (e + f);
y <= a * c + e;
```

Figure 3.18 Incomplete Data Flow based VHDL process description

used to illustrate the basic data flow based scheduling algorithms. The simplest form of scheduling algorithms used in behavioral synthesis is the unconstrained scheduling. It consists of finding a feasible or optimal schedule only obeying the precedence constraints. The most known algorithms are the *As Soon As Possible*(ASAP) and the *As Late As Possible*(ALAP) scheduling. They assign operations to the earliest and the latest possible control step respectively. Generally, these algorithms are used to determine the early and the late execution times of operations and the schedule length. Figure 3.19 shows the ASAP

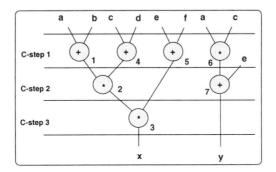

Figure 3.19 Data flow graph and ASAP schedule

schedule of the description of figure 3.18.

Resource Constrained Scheduling

The Unconstrained scheduling represents the basic scheduling problem, but in practice design goals may require other constraints. One of these goals is to limit the chip area. This is done by limiting the number of functional units of each type. The ASAP schedule of figure 3.19 needs three adders. This schedule is not valid if we are constrained to use only one adder and one multiplier. In resource constrained scheduling, the schedule is constructed one operation at a time, while satisfying both the resource and precedence constraints.

The typical heuristic for resource constrained scheduling is *list scheduling* [Hu61]. The list scheduling uses a priority list of ready operations. A ready operation is an unscheduled operation that has all predecessors already scheduled and so can be scheduled in the current control step. During each iteration ready operations are chosen and scheduled in the current control step. A priority function is used to sort the operations and the highest priority operation is chosen to be scheduled in the current control step. Many different priority functions can be

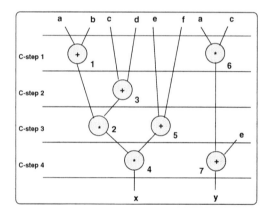

Figure 3.20 List schedule with resource constraint of one multiplier and one adder

used. The simplest one is *mobility* and used in the *SPLICER* system [Pan88]. The mobility of an operation is the difference between its ALAP and ASAP schedules. In figure 3.19, the mobility of operation o_5 is 1, since it can be scheduled at control step 1 or control step 2. The mobility of operation o_1 is 0. The higher is the mobility, the lower is the need to schedule an operation in the current control step. The ASAP schedule needs 3 adders and 3 control steps, but if we schedule the operation 5 in control step 2, we will need only 2 adders. Operations are sorted according to their mobility, so a higher priority is accorded to operations with lower mobility. Take the example shown in figure 3.19 and suppose that the available resources are one multiplier and one adder. During the first iteration, we have four ready operations: o_1, o_4, o_5 and o_6. o_6 is the only multiplication operation, so it is scheduled into the first control step. Since only one adder is available, only one of the three remaining operation can be scheduled in control step 1. o_5 has the highest mobility so, it will be differed. o_1 and o_4 have the same mobility, the one with smaller index is chosen, o_1. For the second iteration only operation o_7 is added to the ready operations, since its only predecessor is scheduled. The new list of ready operations is again sorted and the whole process is repeated. The list schedule of the description of figure 3.18 is shown in figure 3.20 and takes four control steps.

As stated above many priority functions exist. An alternative is to give higher priority to operations with more immediate successors because it makes more of these operations ready than a node with fewer successors. Unfortunately, choosing the best priority function is not a simple task, because it depends on the input description.

Time Constrained Scheduling

Another variant of the basic scheduling problem is the time constrained scheduling. It is of paramount importance for designing typical digital signal processing applications where constraints on the sampling rate of the input data stream have to be met. This is done by constraining the length of the schedule (number of control steps). This constraint may forces some operations to be scheduled into specific control steps. Take the example in figure 3.19 and suppose that the schedule length is limited to three control steps, operations o_1 and o_4 must be scheduled into control step 1, operation o_2 into control step 2, and o_3 into control step 3. These operations are said to be on the *critical path* since we have no choice to schedule them elsewhere.

The most popular method for scheduling under timing constraints is the Force Directed Scheduling(FDS) [PK89]. The goal of this technique is to distribute uniformly operations of the same type into all the available control steps. This uniform distribution results in a higher functional unit usage. The FDS constructs the ASAP and ALAP schedule in order to determine the schedule interval of each operation (range of control steps), for example the schedule interval of operation o_5 is [1,2], where 1 means its earliest execution denoted by E_5 and 2 is its latest execution denoted by L_5. FDS assumes that each operation has a uniform probability of being scheduled into any of the control steps in its schedule interval. Thus, the probability $P_{i,j}$ of an operation o_i to be scheduled into a particular control step s_j such that $E_i \leq j \leq L_i$ is given by : $P_{ij} = 1/(L_i - E_i + 1)$. Figure 3.21 illustrates these probabilities for the example shown in figure 3.19. The operations in the critical path, o_1, o_2, o_3 and o_4 have a probability equal to 1 for being scheduled into their corresponding control steps. The width of each operation box represents the probability of scheduling the corresponding operation into that control step as shown in figure 3.21(a). For example, operation o_5 has a probability of 0.5 of being

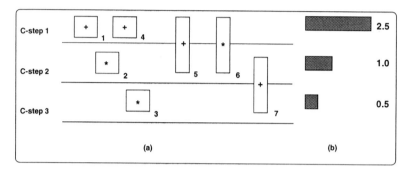

Figure 3.21 (a) probability of scheduling operations into control steps, (b) initial addition distribution graph

scheduled into either control step 1 or control step 2. For each functional unit of type k, a distribution graph is created. It represents the expected operator cost ($FCost_{k,j}$) in each control step j and is given by:

$$FCost_{k,j} = c_k \sum_{i,j \in range(i)} P_{i,j}$$

where c_k is the cost of functional unit of type k, o_i is an operation of type k and $range(i)$ is the schedule interval of o_i. Figure 3.21(b) shows the distribution graph for addition operations in each control step. Assuming a unit adder cost, the expected operator cost is computed as follows:

$$\begin{aligned}
FCost_{add,1} &= P_{1,1} + P_{4,1} + P_{5,1} \\
&= 1 + 1 + 0.5 \\
&= 2.5 \\
FCost_{add,2} &= P_{5,2} + P_{7,2} \\
&= 0.5 + 0.5 \\
&= 1 \\
FCost_{add,3} &= P_{7,3} \\
&= 0.5
\end{aligned}$$

Since functional units can be shared across states, the maximum number of functional units of a type k over all the control steps gives a measure of the

VHDL modeling for behavioral synthesis

total cost of implementing operations of type k. This measure is given by:

$$Cost_k = \lceil \max_j (FCost_{k,j}) \rceil$$

Since the main goal of FDS is efficient sharing of functional units across all the control steps, it attempts to balance the $FCost$ value for each operation type. Consider the effect of scheduling operation o_5 into control step 1 or 2. If this operation is scheduled into control step 1, our example need three adders. However, if o_5 is scheduled into control step 2 (figure 3.22(a), the maximum cost of the adder will be 2, as shown in figure 3.22(b)).

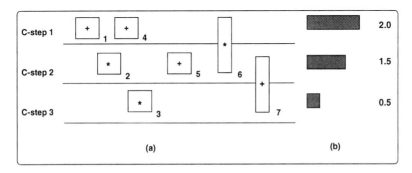

Figure 3.22 (a) probability of scheduling operations into control steps after operation 5 is assigned to control step 2, (b) new addition distribution graph

The FDS starts by creating the distribution graphs for each functional unit, then it computes the expected functional unit cost of scheduling each unscheduled operation into each control step in its schedule interval and assigns an operation to a control step if it results in the minimum cost. The computation of the expected functional unit cost of scheduling an operation into a particular control steps is done by a sum of forces called *direct force* and *indirect force*. A direct force of assigning an operation o_i of type k to a particular control step j is the difference between the expected functional unit cost in j and the average expected functional unit cost over that operation schedule interval and is given by:

$$DForce_{i,k,j} = FCost_{k,j} - \sum_{s \in range_i} \frac{FCost_{k,s}}{L_i - E_i + 1}$$

If we assign an operation to a particular control step, we can affect the schedule intervals of other operations which represent the indirect force of this assignment. The total force of assigning operation o_5 into control step 1 is:

$$Force_{5,add,2} = DForce_{5,add,2}$$

$$= 2.5 - (2.5 + 1)/2$$
$$= 0.75$$

However if we assign o_5 to control step 2, the total force will be:

$$Force_{5,add,2} = DForce_{5,add,2}$$
$$= 1 - (2.5 + 1)/2$$
$$= -0.75$$

It is clear that assigning operation o_5 to control step 2 is the better choice. FDS would choose the schedule producing the largest decrease in force.

Time and Resource Constrained Scheduling

Combining the resource constrained and the time constrained scheduling leads to the time and resource constrained scheduling (TRCS). This means that both the schedule length and the amount of resources are fixed. The goal of TRCS is to find an optimal schedule which respects the time and resource constraints in addition to the precedence constraint.

The most known method to solve this problem is to use mathematical programming (Integer Linear Programming). This solution is guaranteed to produce an optimal schedule. However the optimality comes at the price of the algorithm runtime. Unfortunately, ILPs cannot in general be solved in a polynomial time. Note that ILP methods can solve either the time-constrained scheduling or the resource constrained scheduling. The combination of these problems leads to the TRCS. ILP methods model scheduling as an objective function constrained by a set of inequalities. Let FU_k be the set of functional units available of type k. Let C_k be the cost of such unit and N_k be the number of units performing operation of type k. Let S be the set of control steps. A decision variable x_{ij} is set to 1 if operation i is scheduled into control step j, 0 otherwise. The constraints are formulated as follows:

- Ensure that each operation is scheduled into only one control step in its schedule interval. This leads to the following condition:

$$\sum_{j \in range_i} x_{i,j} = 1$$

for all operations.

- Guarantee that the number of functional units of type k does not exceed K_n in each control step:

$$\sum_{i \in type_k} x_{i,j} \leq K_n, \forall k, j$$

- Ensure the precedence constraint, in other words, for an operation o_j, all its predecessors are scheduled in an earlier control step:

$$\sum_{s \in range_i} (s \times x_{i,s}) - \sum_{t \in range_j} (t \times x_{i,t}) \leq -1$$

for all i,j such that i is an immediate predecessor of j.

Figure 3.23 illustrates the ILP formulation of the scheduling example assuming a unit cost for all the functional units. This schedule has a time constraint of three control steps and a resource constraint of two adders and one multiplier. We can also define the time constrained problem using an ILP by defining an objective function minimizing the cost of functional units usage:

$$\min \sum_{k=1}^{m} C_k \times N_k$$

For the resource constrained scheduling, we can minimize the number of control steps by introducing a dummy operation o_d and adding all appropriate edges ensuring that o_d is scheduled as early as possible. The objective function should be:

$$\min \sum_{s \in range_d} s \times x_{d,s}$$

From a theoretical point of view, ILP algorithms have proven themselves capable of producing optimal solutions and are very powerful. However, they have a major problem which is that the time needed to resolve the set of inequalities grows exponentially with the number of variables, making it unable to schedule large and realistic descriptions. Some heuristics, in order to reduce the size of the ILP-problem, look at a few cycle steps at a time. Zone Scheduling [HH93] divides the set of control steps into two zones such that the problem size of the first zone is small enough to be solved optimally by the ILP approach. Timmer et al [TJ95] propose exact scheduling strategies based on *Graph Bipartie matching* for the TRCS problem. They prove that the existence of a feasible schedule for the TRCS can be decided by finding a correct ordering of operations rather than generating an exact schedule for each operation(ILP) subject to time consuming.

$$\text{Assignment Constraint} \begin{cases} x_{1,1} &= 1 \\ x_{2,2} &= 1 \\ x_{3,3} &= 1 \\ x_{4,1} &= 1 \\ x_{5,1} + x_{5,2} &= 1 \\ x_{6,1} + x_{6,2} &= 1 \\ x_{7,2} + x_{7,3} &= 1 \end{cases}$$

$$\text{Resource Constraint} \begin{cases} x_{1,1} + x_{4,1} + x_{5,1} &\leq A_n \\ x_{5,2} + x_{7,2} &\leq N_a = 2 \\ x_{7,3} &\leq N_a = 2 \\ x_{6,1} &\leq N_m = 1 \\ x_{2,2} + x_{6,2} &\leq N_m = 1 \\ x_{3,3} &\leq M_n = 1 \end{cases}$$

$$\text{Precedence Constraint} \ \{ \ 1.x_{6,1} + 2.x_{6,2} - 2.x_{7,2} - 3.x_{7,3} \ \leq \ -1$$

$$\text{Solution} \begin{cases} x_{1,1} &= 1 \\ x_{2,2} &= 1 \\ x_{3,3} &= 1 \\ x_{4,1} &= 1 \\ x_{5,2} &= 1 \\ x_{6,1} &= 1 \\ x_{7,2} &= 1 \end{cases}$$

Figure 3.23 ILP scheduling as a TRCS

3.3.2 Control flow Based Scheduling

In contrast to DFBS, CFBS use a control flow graph in its strict sense as the underlying internal representation for scheduling. Using data flow graphs is suitable for the synthesis of systems which repeatedly perform a series of operations on an infinite data stream and in which a suitable number of operations can be executed in parallel. This assumption is not applicable to control-dominated systems in which the control sequence is based on external conditions and where the algorithmic description mainly contains control constructs and only few arithmetic operations.

```
entity modn is                              -- algorithm
    port( start : in bit;                   a := ai; b := bi; n := ni;      --1
         ai   : in integer;                 s := 0; i := 0;                 --2
         bi   : in integer;                 while(i <= 15) loop             --3
         ni   : in integer;                     if(odd(b))                  --4
         si   : out integer);                       then s:=s+a;            --5
end modn;                                           if(s>n)                 --6
architecture behavior of modn is                        then s:=s-n;        --7
begin                                               end if;
    process                                     end if;
        variable a : integer;                   i := i + 1;                 --8
        variable b : integer;                   b := b div 2;               --9
        variable n : integer;                   a := a * 2;                 --10
        variable s : integer;                   if (a>n)                    --11
        variable i : integer;                       then a:=a-n;            --12
    begin                                       end if;
        -- algorithm                        end loop;
    end process;                            si <= s;                        --13
end behavior;
```

Figure 3.24 VHDL description of $ab \bmod n$ function

Similarly, conditional branches cause optional program paths. Thus, the number of control steps may vary from one branch to another. However it is fixed for each branch. The DFBS approaches assume that all branches have the same length as the longest path.

Since CFBS are mainly sequential-based approaches, we have to define scheduled paths from the CFG and construct the FSM corresponding to the resulting schedule. Scheduled Paths implicitly represent a function to be executed in one clock cycle at a given time from the whole execution time of the description modeled by the CFG. In other words, the CFG will be partitioned in a sequence of Scheduled Paths. In the following sections, we will use as example the CFG representing the VHDL description of the computation of the function **ab mod n** [Tri85], given $0 \leq a, b \leq n$, and $\lg(n) \leq 15$ as shown in figures 3.24 and 3.25.

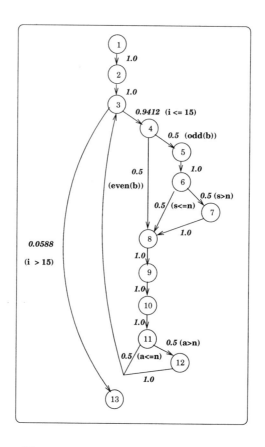

Figure 3.25 CFG of the *ab mod n* function.

Definition 3.1 (Scheduled Path) *Let $G = (V, E)$ be a CFG, let $(C_1, \ldots, C_r)$ be a set of constraints. A Scheduled Path SP is a sequence of nodes $(v_1, \ldots, v_n)$, extracted from the CFG and satisfying all the constraints. If $(C_1, \ldots, C_r)$, the set of constraints, $\prec C_{k(k=1,\ldots,r)}(v_i, v_j) = 1$ this means that the constraint C_k is not satisfied between nodes v_i and v_j.*

Therefore, a Scheduled Path has to satisfy the following property:

Property 3.1 *for all $v_i, v_j \in SP$, for all $C_{k(k=1,\ldots,r)}$, $\prec C_k(v_i, v_j) = 0$.*

We define for each Scheduled Path three parameters:

- a header which is the first node of that path v_1, we note $Header(SP)$.
- a successor, which is the immediate successor of the last node in SP, and we note $Succ(P)$.
- a condition $Cond_{SP}$

The condition enabling the execution of SP, $Cond_P$ is derived by logically ANDing the conditions on the control flow edges leading from Header(SP) to Succ(SP). More formally:

Property 3.2 *Let $SP = (v_i, \ldots, v_j)$, then the condition $Cond_{SP}$ enabling the execution of SP is derived as:*

- $Cond_P(SP) = \wedge_{(i=1,\ldots,n-1),(j=i+1)} Cond(v_i, v_j) \wedge Cond(v_n, Succ(SP))$

Definition 3.2 (CFG Headers) *For each CFG, we define a set of Headers $H(G)$, which represents the set of all the headers of Scheduled Paths.*

$$H(G) = \{v_i \in V | \exists SP \subset V, v_i = Header(SP)\}$$

Scheduling produces a finite state machine FSM. The FSM is modeled either by a Moore or a Mealy automaton.

Definition 3.3 (control flow Scheduling) *Let $\Psi = (S, I, O, f, g)$ be a Mealy FSM, a schedule of a CFG $G = (V, E)$ is a mapping:*

$$\sigma: \begin{array}{rcl} H(G) & \longrightarrow & S \\ v_i & \longmapsto & s_i \end{array} \quad \text{Where:}$$

- $f(s_i, I_k)_{(k=1,\ldots,M)} = \sigma(v_j)$ such that $v_j = Succ(SP_{I_k})$.

The control flow based scheduling [RJ95a] consists of merging all the scheduled paths with the same header into the same state, and a transition is made between two states S_i, represented by v_i, and S_j, represented by v_j, if and only if there is a scheduled path having as Header, v_i, and as successor, $Succ(SP) = v_j$. Then if the machine is currently in cycle step s_i and it is presented an input condition $Cond_P(SP)$, then it will change its control step to $f(s_j, Cond_P(SP))$, and output the result of the execution of all the operations in SP.

Cost Function

Data-dependent loops and loops with non-static bounds in control flow graphs introduce a major problem for the cost function of the scheduling result. This is due to the unknown number of iterations for each loop. The number of states or transitions generated by the schedule does not reflect the real total execution time of an algorithm. The right way to evaluate these algorithms is to define a new metric representing the expected number of clock cycles of a schedule [BDB94]. This metric includes:

- *Branch Probability:* association of a branch probability to each edge in the control flow graph. These probabilities are computed by simulating the given behavioral description with a large set of different possible inputs.

- *Path Probability:* Let $P = (v_1, \ldots, v_n)$, $v_m = Succ(P)$, the probability of executing such a path is:

$$Prob(P) = p(v_1, v_2) \times p(v_2, v_3) \times \ldots \times p(v_{n-1}, v_n) \times p(v_n, v_m) \quad (3.1)$$

- *State Transition Probability:* Let $(P_i, \ldots, P_j)$ be the paths having S_k as entry state, and S_l as destination state. The probability of state transition from S_k to S_l is:

$$p_s(k, l) = Prob(P_i) + Prob(P_{i+1}) + \ldots + Prob(P_j) \quad (3.2)$$

- *Expected number of Clock Cycles of a Schedule:* Let $S = (S_1, \ldots, S_n)$ be the set of states resulting from the schedule. The expected number of clock cycles needed to execute the corresponding input behavioral description is:

$$X_{sch} = \sum_{i=1}^{n} X_i \quad (3.3)$$

where X_i is a random variable representing the expected number of times the state S_i is executed during an execution of the behavioral description.

Computing X_i, $\forall i \in (1, \ldots, n)$ is equivalent to resolving the following set of linear equations:

$$X_1 = 1 \quad (3.4)$$

$$X_i = \sum (X_j \cdot p_s(j, i)) \quad (3.5)$$

$\forall j$ such that $\exists$ a path P from S_j to S_i.

A behavioral description is basically control dominated when it contains a few arithmetic operations and many control constructs. In addition, the control sequence is based on external conditions. The scheduling techniques which operate in a data flow representation are not suited to schedule such a description. Thus, we need algorithms which take into account such properties. Path-based scheduling(PBS) [Cam91, BCP91, CB90] resolves all these problems by considering all program paths and by exploiting optimization potential which concerns state assignment, especially in conditional branches. To ease path extraction and analysis, PBS uses a CFG structure. In this section, we will present scheduling techniques which are suited to efficiently schedule control flow dominated descriptions.

As Fast As Possible scheduling

This approach keeps its originality compared to the others (DFBS) by scheduling all the possible paths independently in an optimal fashion, it then minimizes the path length of a schedule globally. This requires scheduling one operation into different control steps depending on the path, a capability which is missing in DFBS approaches. The basic idea of this approach is to minimize the number of control states under given constraints. This is done for each possible path separately. Paths are computed using a depth-first traversal of the control flow graph. We have to note that the CFG is made acyclic by removing feed-back edges in loops. For instance, in figure 3.25, (v_{11}, v_3) will be removed. The second step is the computation of all paths, starting from the first node of the Acyclic CFG and then all first nodes of loops, since loops are repetitions of sequences of operations. The next step is to compute the constraints for each path. For the sake of simplicity, we consider data dependencies as the only constraint. Constraints are represented as intervals. An interval is a sequence of operations where a new control step must be started at one of these operations. For example, the interval (v_6, v_7) is derived because there is a data dependency between node v_5 and node v_7, so a new control step must be started at node v_6 or v_7 as shown in figure 3.26. The interval graph for the set of constraints of each path is constructed. Each node in the interval graph corresponds to an interval and an edge between two nodes means that the two corresponding intervals overlap. A minimum clique covering is computed consisting of the minimal number of cliques so that each node is only in one clique. Each clique will correspond to a set of operations at which a state must be started. States are ordered along one path. In figure 3.26, three cliques exist, meaning that this path will be executed using four states, three due to the cliques and one

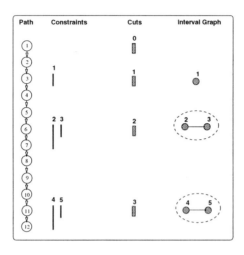

Figure 3.26 Scheduling one path using AFAP scheduling

starting with the node v_1. Now, we have to find the minimum number of states for all paths, by combining the schedules for all paths. Cuts in different paths may overlap at common nodes. This problem can be formulated as that of the constraints. In other words, cuts represent nodes and an edge between two nodes means that the two corresponding cuts overlap. Paths are divided by their cuts into path segments called *Scheduled Paths* as given in definition 3-1. In figure 3.26, the path $(1, 2, 3, 4, 5, 6, 7, 8, 9, 10, 11, 12)$ will be divided into the scheduled paths $SP_1 = (1, 2)$, $SP_2 = (3, 4, 5)$, $SP_3 = (6, 7, 8, 9, 10)$, $SP_4 = (11, 12)$ as shown in figure 3.27. Each path has a header and a successor as given in definition 3-1. The number of headers corresponds to the number of states in the FSM as given in definition 3-3. Thus, all the scheduled paths with the same header will be merged into the same state. That is why this scheduling is defined as a mapping σ from the set of headers to the set of states. A transition is made between two states $S_i = \sigma(v_i)$ and $S_j = \sigma(v_j)$ if there is a scheduled path having as header v_i and successor v_j.

Now, we will see how to evaluate this algorithm using the cost function model defined previously. First of all, we have to review some assumptions needed to validate this model. The CFG is modeled as a finite discrete-time homogeneous Markov chain, meaning that the probability of going from an operation represented by a node v_i to that represented by node v_j is independent of the operations that have been executed before the operation represented by

VHDL modeling for behavioral synthesis

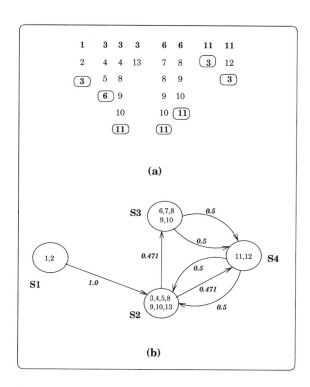

Figure 3.27 (a) Scheduled paths. (b) FSM for AFAP scheduling

v_i. So, each edge (v_i, v_j) in the control flow graph has a branch probability p_{ij} meaning that v_j is executed with a probability p_{ij}, if v_i is executed. In the example of figure 3.25, the branch probability of (v_3, v_4) is $p_{34} = 0.9412$. The way of computing such probabilities can be found in [BDB94]. The next step is to compute the probability, *Path Probability*, for each scheduled path. Consider the scheduled path $SP = (3, 4, 5)$, where $Header(SP) = 3$ and $Succ(SP) = 6$. Using formula 1.3 in section 3.3.2, the path probability of SP is $Prob(SP) = p_{34} \times p_{45} \times p_{56} = 0.471$. Since all the path probabilities are computed we have to calculate the state transition probabilities. We want to calculate for example $p_s(3, 4)$. State S_4 can be reached from state S_3 along two possible scheduled paths: $SP_1 = (6, 7, 8, 9, 10)$ and $SP_2 = (6, 8, 9, 10)$. Using formula 3.2, $p_s(3, 4) = Prob(SP_1) + Prob(SP_2) = 0.5 + 0.5 = 1$. Let X_1, X_2, X_3, X_4 be the random variables representing the expected number of times states S_1, S_2, S_3, S_4 are executed during the total execution of the behavioral description modeled by the CFG. Computing the expected execution time of the AFAP schedule is equivalent to resolving the set of linear equations, us-

ing formulae 3.4 and 3.5:
$X_1 = 1$
$X_2 = X_1 + X_4$
$X_3 = 0.471 X_2$
$X_4 = 0.471 X_2 + X_3$

The solution is, $\{X_1, X_2, X_3, X_4\} = \{1, 17.0, 8.01, 16.03\}$. Finally, from (3), $X_{sch} = 1 + 17.0 + 8.01 + 16.03 = 42.04$ clock cycles.

Dynamic loop scheduling

In contrast to AFAP scheduling, DLS [ORJ93, ROJ94] keeps all loop feedback edges in the control flow graph and interrupts the generation of paths *on the fly*. In other words, the generation of a current path is stopped only if a constraint is violated. This reduces the complexity of the number of generated paths.

Starting from the first node of the CFG, all the paths are calculated. The first path starts with the first node in the CFG. Successor nodes are sequentially appended to the path until one of the constraints is violated. The path generation algorithm is outlined in figure 3.28. The algorithm is invoked by calling

> **SearchPath(n)**
> **Begin**
>   Add n to the list of Headers;
>   Initialize a new path, P(n);
>   For Each x $\in$ n.successors BuildPath(P(n),x)
> **End SearchPath;**
> **BuildPath(P(n),x)**
>   **If** Constraint_violated(P(n),x)
>   P(n).successor := x;
>   If x is not already a Header then SearchPath(x);
>   **Else** Append x to P(n);
>   For Each y $\in$ x.successors BuildPath(P(n),y);
>   **End If**
> **End BuildPath;**

Figure 3.28 Path generation algorithm for DLS

VHDL modeling for behavioral synthesis

the function *SearchPath(n)*, passing the first node of the CFG as parameter. This initializes a new path, *P(n)*, with *n* as the *header*. For each successor of *n*, a new path is begun by calling the algorithm *BuildPath(P(n),x)*, where *x* is a successor of *n*. Each time one of the constraints is violated, the current path is terminated and a new one begins. The first node of the new path becomes the successor of the current path, and *SearchPath()* is once more called to initialize a new set of paths. The paths for the example of figure 3.25 are shown in figure 3.29. We have to note that the only constraint considered in this example is data dependency. The rectangular box at the bottom of each path contains the successor node of that path. Once the paths are generated, the FSM is constructed with respect to definition 3-3. The FSM for DLS is shown in figure 3.30.

1	3	3	3	6	6	11	11	11	11	11	11	5	10
2	4	4	13	7	8	12	12	12	3	3	3	(6)	(11)
(3)	5	8		8	9	3	3	3	4	4	13		
	(6)	9		9	10	4	4	13	5	8			
		10		10	(11)	(5)	8		(6)	9			
		(11)		(11)			9			10			
							(10)			(11)			

Figure 3.29 Paths and successors for the Dynamic Loop Scheduling

DLS differs from AFAP in two major areas. Firstly, DLS keeps all loop feedback edges in the control flow graph, thereby allowing the *parallel execution* of parts of *successive loop iterations*. This is validated for the paths $SP_7 =< 11, 12, 3, 4, 8, 9 >$, $SP_8 =< 11, 12, 3, 4 >$, $SP_9 =< 11, 3, 4, 5 >$, $SP_{10} =< 11, 3, 4, 8, 9, 10 >$ shown in figure 3.29, where part of operations of iteration i of the loop can be executed in parallel with a part of iteration $i+1$. Consider the execution of the sequence $< 3, 4, 8, 9, 10, 11, 3, 4, 8, 9, 10, 11, 12, 3, 13 >$ of the CFG of figure 3.25. DLS executes the sequence by executing the states $< S_2, S_4, S_4 >$, which requires three clock cycles. To execute the same sequence, the AFAP schedule executes the states $< S_2, S_4, S_2, S_4, S_2 >$, requiring 5 clock cycles. The fact that we can pipeline different iterations of the loop means that the execution time of an algorithm containing loops scheduled with DLS will be faster than for AFAP. We can see that the expected number of clock cycles needed to execute the entire description shown in figure 3.25 using DLS, is 33.5. The second major difference is that DLS interrupts the generation of paths *on the*

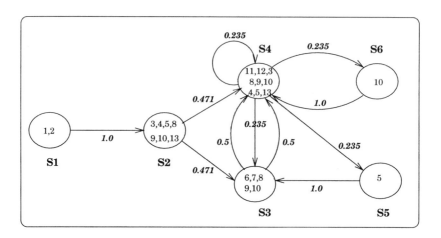

Figure 3.30 FSM for the Dynamic Loop Scheduling

fly. One of the major limitations of the AFAP schedule is that, as the CFG becomes complex, the number of paths generated quickly increases making the AFAP approach very cumbersome for large CFGs. The *on the fly* generation of paths dramatically reduces the number of paths.

Pipeline path-based scheduling

When dealing with loop intensive specifications, AFAP approach tends to be sub-optimal when the input description contains many loops. The problem is related to the fact that, in this approach, all loop feedback edges are broken and thus no advantage can be taken of the fact that different loop iterations can be pipelined, implying potential parallelism beyond loop boundaries. DLS attempts to overcome this problem by leaving loop feedback edges intact. However, this approach is rather simplistic as it only considers one iteration. In addition, it does not cut the generated paths in an optimal way as it uses an As Late As Possible (ALAP) scheduling technique to do this. Nevertheless, results published for these algorithms show that by treating loops more efficiently, improvements on the original AFAP approach can be made, even on taking into consideration the fact that the path cuts are not optimal.

The main idea of the Pipeline Path-based Scheduler(PPS) [RJ95b] is to benefit from the advantages offered by both AFAP and DLS:

VHDL modeling for behavioral synthesis

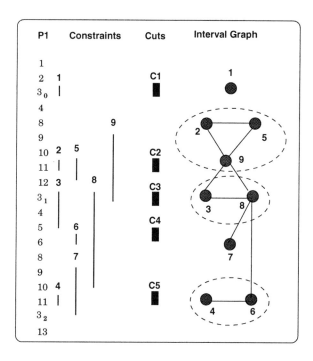

Figure 3.31 Scheduling one path using PPS

- Minimizing the number of cuts and then the number of states similar to AFAP which results in an optimized controller implementing the FSM produced by the scheduler.

- Pipelining the execution of loops to minimize the number of clock cycles needed for the execution of a required description.

To achieve these goals, PPS considers that a given loop will be executed 0, 1, 2 or more times. If there are no loops, the paths are generated in accordance with the AFAP approach. If on the other hand loops exist, paths will be generated assuming the loop executes 0 times, once or twice. This technique is not unlike that of loop unrolling [GVM89]. Unrolling loops twice allows us to detect any inter-dependencies that may exist between different iterations of the same loop.

Minimizing states

In figure 3.31, we show a PPS schedule of the path $P = <1, 2, 3_0, 4, 8, 9, 10, 11, 12, 3_1, 4, 5, 6, 8, 9, 10, 11, 3_2, 13>$, that has been generated by unrolling the loop twice. As PPS uses the minimum clique covering technique to compute the constraints, we can see that compared to DLS, we do not need to cut the path at node 5 and at node 10 since we have already a cut at node 3. When using DLS, the cut at node 3 cannot be observed, because DLS uses an ALAP fashion for cutting paths and this results in two more states as it can be seen from figures 3.30 and 3.32 which represent the FSMs for both schedules.

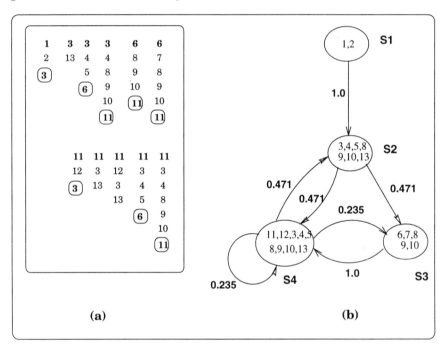

Figure 3.32 (a) Scheduled Paths for PPS, (b)FSM for PPS.

Minimizing the execution time

Now, assume that we have to execute the sequence $P = <3, 4, 8, 9, 10, 11, 12, 3, 4, 5, 6, 8, 9, 10, 11, 3, 13>$. The AFAP schedule goes through the states $<S2, S4, S2, S3, S4, S2>$ to execute this sequence. The PPS goes through the states $<S2, S4, S2, S3, S4>$. While AFAP schedule requires 6 clock cycles, PPS needs only 5 clock cycles to execute the same sequence. For the same execution sequence, DLS goes through states $< S2, S4, S5, S3, S4 >$ which require 5 clock cycles as PPS, but DLS needs another state $S5$ to perform this execution. This advantage can be validated by the expected number of clock cycles required by PPS to execute the entire description of figure 3.25, evaluated to 32.55.

Discussions

Algorithm	State	Transition	Clock_Cycle
AFAP	4	8	42.04
DLS	6	14	33.5
PPS	**4**	11	**32.5**

Table 3.2 Results for the *ab mod n* algorithm.

Table 3.2 shows the different results for the three approaches, AFAP, DLS and PPS. *State* means the number of states, *Transition* the number of transitions and *Clock_Cycle* the expected number of clock cycles needed to execute the algorithm. When looking at the FSMs produced by AFAP and DLS, we see that the one produced by DLS is more complex and contains more states than that of AFAP. Nevertheless, using DLS we kept 8 clock cycles for the entire execution of the *ab mod n* algorithm. This is due to the fact that AFAP will always take at least two states to execute the loop. However, under certain conditions, DLS can take one state (S4). By combining the advantages of both approaches, PPS minimizes the number of states and transitions. Moreover, by unrolling loops it can minimize the number of clock cycles.

3.3.3 Scheduling for programmable architecture

Most existing techniques in high-level synthesis assume a simple controller in the form of a hardwired FSM and can only execute one specific application program. This controller model is characterized by the fact that the control logic is synthesized after scheduling when the state transition graph is known. Thus, before scheduling, we do not know how many transitions are issued from any state and therefore, the state transition graph may contain many branches. In contrast, a programmable controller consists of two major parts: a program

ROM, E-PROM or RAM to store the instructions and a sequencer block, which controls the execution sequence of the program. The main characteristic of this controller is that the sequencer block and the instruction decoder are fixed in advance. This implies that the number and type of branches are limited by the available control flow instructions in the instruction set. To be able to generate programmable controllers, we have to take into account such characteristics during scheduling.

To generate programmable architectures, two constraints must be adhered to:

- All conditions must be executed in the data path and will not be synthesized by a dedicated logic block, as in the case of the hardwired solution.
- The number of multiple branches must be reduced with respect to the available control flow instructions in the instruction set.

[KGM95] presents a unified scheduling model for high level synthesis and code generation. This algorithm uses a CDFG model extended by three special nodes $CJUMP$, $MERGE$ and $JUMP$ to represent a conditional branch. The $CJUMP$ node starts the conditional branch by evaluating the values of conditions. The $MERGE$ nodes mark the definition of signals based on different values produced in each conditional path. *Weak control dependencies* edges are introduced to denote dependencies violation. They are added between every $CJUMP$ node and all operations inside the conditional branch started by this $CJUMP$ and ended by the corresponding $JUMP$. Figure 3.33 shows a CDFG model of a conditional branch as used in [KGM95]. Virtual resources are introduced in the controller to execute the $CJUMP$ and $JUMP$ operations called $CJUMP$ operator respectively $JUMP$ operator. Programmable processors contain conditional branch instructions. The $CJUMP$ operations will be mapped onto one of these control flow instructions depending on the condition flag resulting from the CJUMP evaluation. Depending on the number of conditional branches allowed by the programmable controller, the number of $CJUMP$ and $JUMP$ nodes can be limited. For example, in the case of DSP processors, only two-way conditional branch and unconditional branch are available, so only one $CJUMP$ operator will be used. The scheduling technique used in [KGM95] is a list-based scheduler with the following strategies:

- The execution condition of an operation is updated whenever a weak dependency to that operation is violated.

VHDL modeling for behavioral synthesis 119

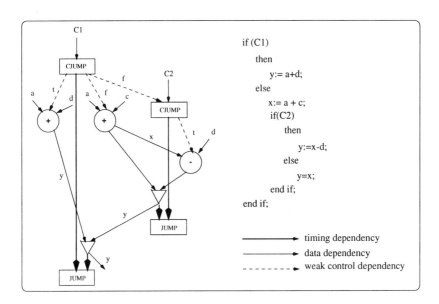

Figure 3.33 CDFG model of a conditional branch

- The priorities of operations that violate the weak dependencies are lowered such that they will not compete too heavily for resources with other operations which retain their execution condition.

- To allow resource sharing, the priorities of two mutually exclusive operations that share the same resources are increased by the scheduler.

[RBL+96, RJ96b] presents an approach called *Performance Analysis based Scheduling* to solve the problem of scheduling for programmable controllers. This technique is tuned for control dominated applications. It transforms a complex FSM into an equivalent one which captures the characteristics of a programmable controller, for example, the available control-flow instructions. This transformation is carried out while minimizing the overall execution time.

This approach combines several algorithms and is done in two independent tasks:

- Execution of a control-flow dominated scheduling algorithm without taking into consideration the additional constraints mentioned above.

- Transformation of the resulting FSM in order to satisfy the aforementioned additional constraints. This task can itself be broken into three subtasks:
  - Transform each multiple branch into an equivalent set of binary branches.
  - Find the optimal ordering of these binary branches using the performance analysis technique described in section 3.3.2.
  - Schedule conditions using a classical list scheduling approach.

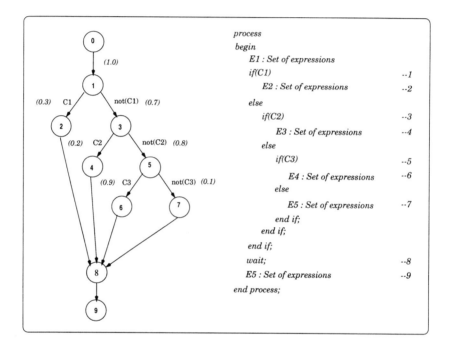

Figure 3.34 Control-flow graph, VHDL process description

To present this approach, let us take the CFG representation shown in figure 3.34. A control-flow based scheduling of this CFG results in the paths and FSM shown in figure 3.35.

We would like to map this FSM into a programmable one in which the control flow instructions are limited to binary conditional and unconditional branches. This transformation must be carried out while trying to minimize the total execution time. This implies the necessity of introducing certain intermediate states. Within these limits, we notice that priority must be given to the branches having the highest probability of execution as proved in [RJ96b]. In

VHDL modeling for behavioral synthesis 121

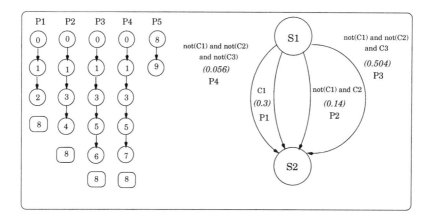

Figure 3.35 Paths and FSM schedule example

our example, the only state having multiple branches is state $S1$. Our starting point is $S1$. We select the branch emanating from $S1$ which has the highest probability of execution. In this example, this corresponds to the execution of path $P3$. We then proceed by inserting an intermediate state $S1_1$. The insertion of state $S1_1$ is required as it will enable us to restart from an initial state while preserving the binary branching process (no more than two branches per state). The same therefore, can be said of the following branching. That is, we select the branch from state $S1_1$ that has the highest probability of execution. Another intermediate state has to be inserted $S1_2$ to preserve the binary branching constraint. At this point, there remains only two branches corresponding to the execution of paths $P2$ and $P4$ and they must emanate from $S1_2$, as shown in figure 3.36.

Since the optimal ordering of these binary branches is found, the next step is to perform the evaluation of all conditions of a multiple branch simultaneously. From a scheduling point of view, the best way to represent these conditions is in the form of a data flow graph. This has the added benefit of facilitating the execution of typical optimization techniques such as redundant sub-expression elimination. In addition, the least probable condition will not have to be executed as this condition is simply the negation of the penultimate condition. For example, from figure 3.37, the condition F3 will not have to be executed as it will be replaced by the condition not(F2) as shown in figure 3.36. A list scheduling algorithm [Hu61] is executed on the resulting data flow graph so as to minimize the overall execution time.

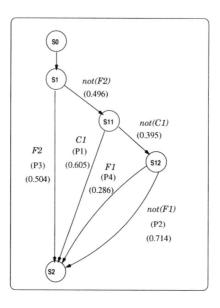

Figure 3.36 Performance analysis based scheduling result

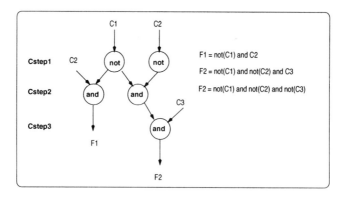

Figure 3.37 DFG and Schedule of Conditions

To show the efficiency of this technique in terms of total execution time, we have to compute the latter by resolving the set of linear equations:
$X_0 = 1$
$X_1 = 1$
$X_{11} = 0.496 X_1 = 0.496$
$X_{12} = 0.395 X_{11} = 0.196$

$X_1 = 0.504X_1 + 0.605X_{11} + X_{12} = 1$

The total execution time is $Xsched = 3.692$ control steps. Without reordering the transitions, we would reach a solution (having the same number of states and transitions) in which the total execution time would take 4.26 control steps. The schedule presented in [KGM95] would take 5 control steps and need 8 states to execute.

Note that this example was chosen simply to illustrate the basic concepts. For more complex examples (containing many loops and branching instructions) such as real time systems, the gap between the presented solutions is much more significant.

The performance based technique gains from the fact that it can benefit from the level of parallelism generated by a control-flow based schedule and is therefore not obliged to add cuts at each branch. In addition, it takes advantage of some well-known compilation techniques to optimize the evaluation of the conditions. Performance analysis also allows us to generate an optimal solution in terms of the execution time of the intermediate states.

3.4 SUMMARY

Behavioral synthesis performs several complex transformations of the behavioral specification before the generation of an RTL model. These transformations make difficult to predict the results of behavioral synthesis. This chapter discussed three issues aimed to ease the understanding of the results of behavioral synthesis: VHDL interpretation, VHDL execution models and scheduling VHDL descriptions.

Each VHDL construct may be interpreted in different ways according to the behavioral synthesis tool strategies. Most tools start with an entity/architecture pair and deal with a single VHDL process. Conditional statements may be assigned t the controller or to the datapath.

Several transformation schemes may be applied to loops: unrolling, folding, pipelining, ... Procedure may be expanded inline, mapped onto components or interpreted like operations executed on coprocessors. Variables and signals may be mapped on wires or storage units using one of the different existing approaches.

Six execution modes are defined for VHDL execution. These are classified according to the scheduling of VHDL execution threads (code executed between two successive wait statements during simulation) and the scheduling of I/O operations.

Scheduling is one of the main added value of behavioral synthesis. However there are several methods and heuristics to perform this task. Scheduling algorithms can be classified into two major classes: data flow based scheduling (DFBS) and control flow based scheduling (CFBS). DFBS is suitable for data oriented applications acting on regular streams where the behavior is specified as periodic operations. CFBS is suitable for control flow applications where the behavior may include data dependent computations acting on complex data structures. Several scheduling algorithms exist for both DFBS and CFBS.

4

BEHAVIORAL VHDL DESCRIPTION STYLES FOR DESIGN REUSE

This chapter deals with VHDL modeling style for modular specification and design reuse at the behavioral level. Modular specification allows to decompose a large design into smaller pieces which are easier to handle. Design reuse implies to be able to reuse existing components as black boxes during behavioral synthesis. These two concepts, when combined, are the basis of what is called structured design methodologies. These methods may be applied for both manual design and behavioral synthesis.

After an introduction to design reuse and structured design method, specific issues related to the behavioral level will be discussed. This mainly includes the concepts of behavioral component and component abstraction for behavioral synthesis. Section 4.3 deals with modular design and section 4.4 provides guidelines for behavioral VHDL description style allowing design reuse and modular design. Section 4.5 deals with object oriented VHDL which may be the most natural writing style for design reuse and modular specification.

4.1 DESIGN REUSE

The bottleneck nowadays in system design is the designers' productivity (as stated in chapter 1). In order to overcome this, it is necessary to exploit reuse principles. This concept has been applied in several domains; for instance, solving mathematical problems consists in bringing them to problems with known solutions, i.e. so that existing theorems, lemmas or principles may be applied. Therefore the solution to master high complexity is to (1) identify existing elements before (2) reusing existing elements.

Reuse is a common problem to hardware and software design. For both of them, designers have got recourse to the reuse of existing elements. Software reuse is defined as the use of existing software knowledge or artifacts to build new software artifacts [FF95]. Similarly, hardware reuse exploits existing items. While software reuse is broadly defined [FI94] hardware reuse practice is up to now limited to register transfer and lower level designs.

Design reuse allows tremendous gains in the development and the verification time. However design reuse imposes specific methodologies based on modular design. These are called structured design methodologies and can be applied at different abstraction levels in order to master complexity. Structured design methodologies are aimed at modular decomposition of the design in order to allow the separate design of different modules. In addition this allows the reuse of existing components and already designed modules in order to build larger ones.

Design reuse at the behavioral level consists of the reuse of blocks of any degree of complexity, including existing blocks, RTL or logic descriptions and even blocks obtained from previous behavioral synthesis sessions. In fact design reuse can easily be combined with all existing synthesis tools at the behavioral, RT and even lower levels if the blocks to be reused are used as black boxes or rigid macro-blocks. Such a design situation is illustrated in figure 4.1. Starting from a behavioral specification composed of a main algorithm and a set of component instances, the synthesis is run over the main algorithm only. The behavioral synthesis session will generate a main controller and an interface between the latter and the components instanced. The architecture corresponding to the required circuit can be obtained by recombining the structural part regrouping the black boxes to the synthesized blocks.

The above defined solution is simple and can be applied whatever the level of abstraction. In fact, it is currently being used at the RT level. However the architecture generated presents some limitations. Indeed the macro-blocks are considered as black boxes and no information about the functions they execute is given. Therefore no resource sharing can be done between any of the macro-blocks and the synthesized algorithm. However it is often the case that these black boxes need to be used as coprocessors in order to execute operations of the behavioral description. The separation between the behavioral description and the components will induce communication overhead due to timing and synchronization constraints. With such a scheme, all operations executed on these components will require more than one clock cycle. In order to use a simple adder located in one of the components, at least 2 cycles are needed even if the adder can compute in less than one clock cycle. As the data transfers

Behavioral VHDL description styles for design reuse

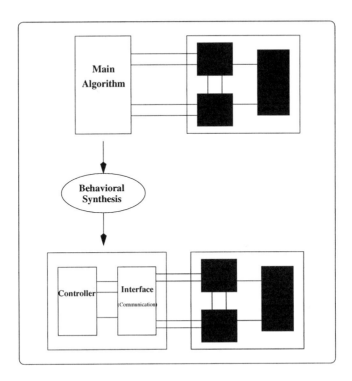

Figure 4.1 Design reuse during behavioral synthesis

are defined within the controller with a granularity not smaller than the clock cycle, the results of the adder can be recovered only one clock cycle after the moment the input values were placed.

An alternate architecture which allows to overcome these constraints can be obtained by using these black boxes as functional units in the behavioral description and during behavioral synthesis. In this chapter we will restrict the analysis to synchronous systems organized as a top controller and a datapath that may include complex functional units acting as coprocessors. This architecture corresponds to the FSMC model (section 2.1) which is the most general scheme allowed by behavioral synthesis. In this second solution, it is necessary to provide the synthesis with some additional informations about the operations and resources available within these black boxes. This design flow can be illustrated by figure 4.2.

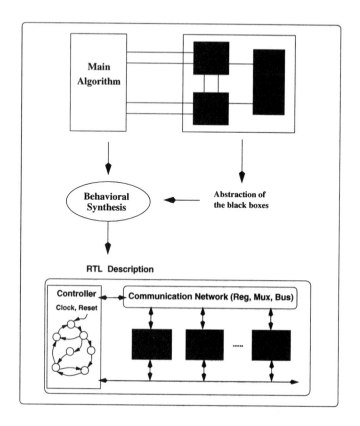

Figure 4.2 Design reuse

Design reuse constraints both the design of the components to be reused and the way the applications have to be constructed starting from the reusable components, however it presents tremendous gains in the development and verification time [Sta94].

4.1.1 Structured design methodology

The structured design methodology for VLSI was introduced first by Mead and Conway [MC80]. This methodology has two main concepts: hierarchy and regularity [TRLG81], thereby leading to modularity. Hierarchies are used for decomposing a complex design into sub-parts that are more manageable. Regularity is aimed to maximize the reuse of already designed components and

Behavioral VHDL description styles for design reuse 129

subsystems. This kind of design methodology has proven its efficiency in several other domains of system design [Seq83].

Each time a new design level is adopted, the design community starts looking for structured design methodologies acting at that level. Structured design methodologies have been adopted since the starting of the domain of VLSI design [MC80]. This methodology has been applied to the physical level with physical design tools [TRLG81], [Seq86]. It has also been applied to logic and register transfer level design [GC93]. In order to define a structured design methodology acting at the behavioral level, we need to solve two closely related problems:

- How to structure the design in order to allow modular design and synthesis at the behavioral level. This would mean that a behavioral design can be used as a component to build more complex designs.

- The reuse of existing components for the design and synthesis at the behavioral level. These components may correspond to the results of an early architecture design process or to existing components that have been designed using specific design environments. This would mean to allow behavioral synthesis to reuse complex components.

4.1.2 Principle

Structured design methodology is based on the divide-and-conquer principle. Regardless the abstraction level, the structured design methodology for VLSI consists of 3 main steps in the design flow:

- Partitioning; it can be applied on the system specification in order to split the system into simpler subsystems. This step is aimed to structuring the design in order to produce a hierarchical modular decomposition of the initial specification. This leads to the isolation of subsystems that will be designed independently as well as the isolation of sub-functions to be executed on existing subsystems. Proper partitioning allows independence between the design of the different parts. The decomposition is generally guided by structuring rules aimed to hide local design decisions, such that only the interface of each module is visible.

- Module design; each subsystem thus generated can be designed independently using a specific library of components. This step makes use of a functional module library. This may include either standard existing functional units or specific modules that have been designed during previous steps. The implementation details of these modules are hidden.

- Module abstraction for reuse; the abstraction of the subsystem has to be done in order to enable its reuse as a complex library element. This abstraction extracts the necessary information for synthesis, i.e. data exchange protocols, physical informations, ...

The two first aspects are closely inter-related. The modular decomposition may be influenced by the set of already existing components. On the other side, the selection of the components is influenced by the modular decomposition of the design. Accordingly, the power and flexibility of such scheme depend on the range of components that may be used in the library.

For the application of structured design methodology to architectural or behavioral synthesis, system analysis and partitioning allow the identification of subsystems from the top controller as illustrated in figure 4.3. The subsystems may themselves be synthesized at the behavioral level according to their degree of complexity. In this case, the design flow will consist of the behavioral synthesis of the subsystems and their abstraction for reuse before the synthesis of the top design. The RTL result for the overall design includes both the top configuration and the register transfer level description of the subsystems.

As stated above, we will restrict this chapter to synchronous models made of a top controller and a datapath with coprocessors. When a system is organized as a set of concurrent modules using a distributed control scheme, several separate behavioral steps may be needed. Each step will handle a separate module.

4.1.3 Design abstraction for reuse

When such a methodology has to be automated, the main issue is to allow component abstraction and therefore to achieve first description for reuse. The component library may contain:

- Components and subsystems that may be designed using other design methods and tools, and
- Sub-systems that result from an early synthesis process.

Behavioral VHDL description styles for design reuse

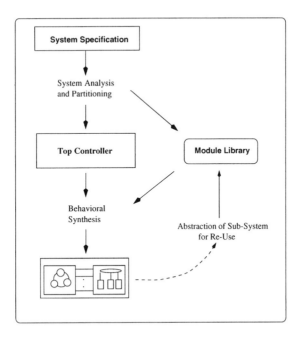

Figure 4.3 Use of structured design methodology

The main pieces of information necessary for the abstraction of a subsystem for its reuse as a functional component are the operations it can execute and their execution scheme. These two factors have to be taken into account when they are designed so that

- the necessary control signals are introduced at the behavioral level to give access to the wanted operators,
- the communication scheme of each of the operators can be fixed.

More details about design modeling for reuse are given in section 4.4.

4.1.4 Hierarchical and modular architecture

As illustrated in figure 4.4, the result of such a design flow is an architecture made up of a control unit (called the top controller) managing a set of functional modules. This architecture may be hierarchical and the composition of each functional module may consist of another control unit and another set of functional units, and so on. The hierarchy ends with functional modules without local control, such as a basic arithmetic unit.

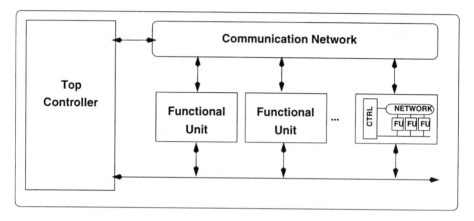

Figure 4.4 Modular architecture

This architecture is still synchronous because of the single top level control unit present in the sub-hierarchy of the architecture. However synchronization is much less rigid than in a flat architecture as each functional module may run independently and in parallel with any other functional module of the architecture. Communication between functional modules and the top controller allows to exchange data according to predefined protocols. We assume that all the modules share a common clock. However each module may have other local clocks.

4.1.5 Resource sharing

Behavioral descriptions consist of a set of operations. These are often complex ones and usually ask for functional modules for their execution. The functional modules can be considered as rigid blocks just as macro-blocks for the register transfer level synthesis. Each of them can execute one or more operations and may be bound to several operation/procedure calls made at different moments. These modules may be handled during behavioral synthesis like functional units executing operations. In the behavioral description a specific operation may be specified as a procedure or function. These black boxes may be allocated, bound and shared in the same way as ALUs or other classical functional modules used by behavioral synthesis tools.

4.2 DESIGN REUSE AT THE BEHAVIORAL LEVEL

At each level of abstraction, different techniques are used for modular decomposition as well as for component specification and reuse.

At the register transfer level, a typical environment will provide methods for modeling generic operators (such as adder, multiplier, ALU, ...) able to execute the basic operations: +, *, -, ... These operators, acting as black boxes in the netlist (input description), are a hybrid between arithmetic operator and library cell. They are the link between the HDL operator and the components of the final library. Each operator corresponds to a unit that may perform one or several functions. Since the global description is given at the clock cycle level (the basic time unit), the execution timing delay of each of these operations cannot exceed a clock cycle. This section extends these concepts to support the behavioral level by introducing a new paradigm called the behavioral component.

4.2.1 Design reuse

At the behavioral level, the available information concerns the system specification and a library of components. This enables two main types of reuse: the reuse of the behavioral description and that of library components.

Reuse of behavioral description

The same behavioral input description may be used to explore the design space and therefore to reach different architectures. Such a design flow is illustrated by figure 4.5(a). Another way to reuse behavioral description is to modify an initial specification in order to build another design that performs a close behavior - for instance extending the functionality of the initial system. This kind of reuse is much easier at the behavioral level than at the lower level. For example, if the initial behavioral description is an existing controller to which we decide to add another addressing mode, the design flow used may be figure 4.5(b). At the behavioral level such a mode will need to add a few lines in the behavioral description. At the RT level, such a modification will induce changes in the datapath (in the case where new resources are needed) and in the controller as well.

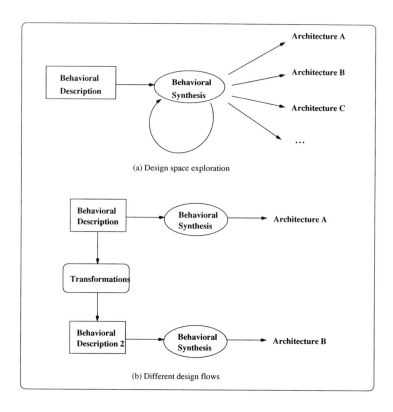

Figure 4.5 Reuse of the behavioral description

A behavioral description may therefore be reused for different architecture generations but may also be reused for different technology mappings. One of the main advantages of behavioral descriptions is that such descriptions are smaller in size and can more easily be read with respect to lower level descriptions. Owing to this high degree of legibility, behavioral descriptions can be updated much faster if there is any evolution within the system specifications.

Reuse of library components

The components may be classified into two main groups: functional modules and basic components. The latters include all communication units such as multiplexers, switches, ... as well as storage units such as register files, registers, ... The main difference between these components and the functional modules is that they are inherent to the synthesis system target architecture.

The architecture and functionality of each of them are known by the system and exploited by it during synthesis. These generally correspond to a set of predefined components. On the other hand, functional modules may consist of simple or standard operators, but also of complex macro-blocks which behave as coprocessors. These are called behavioral components.

4.2.2 Behavioral components

A behavioral component is an entity able to execute a set of operations invoked in the behavioral description. The component acts as a black box linking the behavioral and register transfer levels. The operation(s) executed by the behavioral component may be as simple as predefined operations (+, -, *, ...) or as complex as input/output operations with handshaking or memory access with complex addressing functions. A component may correspond to a design produced by external tools and methods or to a subsystem resulting from an early design session. Complex operations can be invoked through procedure and function calls in the behavioral description. Allowing the use of procedures and functions within a HDL (Hardware Description Language) is a kind of extension of this HDL. This concept is similar to the concept of system function library in programming languages. This way a language is composed of two parts:

- A fixed part which includes the predefined constructs, and
- An exchangeable part which includes a set of procedures and functions that can be used within the language. These need not to be part of the language itself.

Modeling behavioral components as functional units

The concept of behavioral component is a generalization of the functional unit concept. A behavioral component allows the use of existing macro-blocks in the behavioral specification. It is modeled as a coprocessor able to execute a set of operations. It may execute standard operations or new customized operations introduced by the user. A behavioral component can be called from within a behavioral description through procedure and function calls. It can accept and return parameters.

Each functional unit can be specified at four different abstraction levels: the conceptual view, the behavioral view, the implementation view and the high-level synthesis view. Figure 4.6 shows these four views for a memory component that can achieve the 2 operations: *mread* and *mwrite*. From the conceptual point of view, the functional unit is an object that can execute one or several operations which may share some data (M). At the behavioral level, the functional unit is described through the operation that can be called from the behavioral description. These may correspond to standard operations or procedures and functions. The behavioral view in figure 4.6(c) includes the procedures *mread* and *mwrite*. These 2 procedures hide the addressing mode, the data exchange protocol and the internal organization of the memory block. The implementation shows an external view of a possible realization of the functional unit; it thus includes the different connections of the functional unit: inputs, outputs and selection commands (selecting the command to execute).

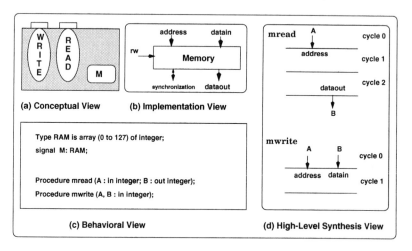

Figure 4.6 The four abstraction levels

In order to be able to handle such units using behavioral synthesis, we need a high-level synthesis view of the functional module to link the behavioral and implementation views. It includes the interface of the functional unit, its call-parameters (corresponding to the operation parameters), the operation set executed by the functional unit as well as the parameter passing protocol for each operation. In the case where the functional module executes complex operations that may require several cycles, this view provides a fixed schedule for the execution of the operations. This is sometimes called an execution template [KLMM95]. This schedule is expressed through static clock cycles.

In order to allow behavioral synthesis to schedule the behavioral description, these complex operations need to have a fixed predictable execution time.

Complex functional units may execute 2 kinds of operations:

- Computation operations. These correspond to functions or procedures which return results that depend only on the input parameters.
- Communication operations. These correspond to procedures used to exchange data with the functional unit. Such operations may return values that result from previous computation performed by the functional unit. This kind of operation is needed in the case of functional units that execute complex algorithms that may include a data dependent computation. In order to execute such an algorithm, the top controller may need the execution of a sequence of operations such as start computation and get result of computation.

In the case where complex algorithms are invoked through a set of communication primitives, the functional unit may use local memory in order to store temporary values. In the general case, a functional unit may have a local memory and the algorithm may use a complex data structure. The local memory of the functional unit may be accessed through specific communicating processes. Figure 4.7 illustrates a functional unit executing a complex algorithm invoked through 2 communicating operations.

- "operation_start" starts the computation of the algorithm. The results of the computation is stored in a local memory called "tmp".
- "result_recovering" returns the results of the computation. Of course this may correspond to a simple value or to a complex data structure.

Linking behavioral operation to behavioral components

According to the characteristics of the application, a set of functional units will be provided before starting the synthesis process. For instance, the use of special purpose hardware units able to increase the computing power of the architecture may be provided. The correspondence between the operations of the behavioral description (standard operators such as + and -, and procedure

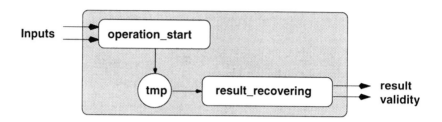

Figure 4.7 Using communication primitives

and function calls) and the functional units is made during the synthesis process. In VHDL, these operations can be modeled as procedure and functions as it will be detailed later. These operations act as a link between the behavioral description and the functional unit library. Initially we may have a library where each operation may be executed on one or more functional units and where each functional unit may execute several operations. We may even have operations that have different execution schemes on different functional units. The number of functional units selected (allocated or instanced) will depend on the parallelism allowed by the initial description. Of course the synthesis process tries to share as much as possible the use of the functional units.

Memory with complex addressing function and specific embedded computation and control can be easily described with this scheme. The addressing functions may be realized by an independent functional unit or may be integrated within the memory unit. In the same way complex I/O units may be used. They are also accessed through function and procedure calls. These may execute complex protocol or data conversion.

Modeling coprocessors as functional units

A coprocessor is a macro (black box) that can execute a set of operations in parallel with the main processor. Therefore coprocessors can be modeled as complex functional units allowing internal memory and undertaking a large set of operations with different timing scheme. Such a modeling lies on the definition of the data transfers to and from the coprocessor. The coprocessor may be a general purpose programmable processor or specific hardware dedicated for the execution of specific operations. The coprocessor may run in an absolute concurrent manner with respect to the rest of the circuit.

The abstraction of a coprocessor as a functional unit consists in:

- Fixing the operation that will be executed on the coprocessor. The execution scheme for each operation should be fixed. For example, in the case of a programmable processor, the algorithm of the operation has to be coded. The execution scheme should take into account communication with the top control, such as the protocol for reading input and writing output.

- Fixing the communication protocol for each operation. This may be a simple procedure call or a complex handshake inducing several communication operations that have to be executed in a given order.

4.3 MODULAR DESIGN

The key step when dealing with complex systems is to structure the design into a set of modules. This process is illustrated in figure 4.8. It consists in

- Identificating sub-behaviors that can be subcontracted to functional modules or coprocessors.
- Identificating or designing a set of functional modules or coprocessors able to execute the functions identified during the first step.
- Reorganizing the specification according to the main decomposition. The main description will be made of a behavioral description and a set of functional modules.

This will be explained through an example of PID.

A PID controller usually applies a control function to an analog input and generates an analog output. This kind of device is generally implemented as an analog device. The use of a digital solution allows to have an integrated solution. The control loop making call to the PID is shown in figure 4.9. The controller device performs the 3 functions: speed control, current control and communication with a host computer. More details on this example are given in chapter 6.

The PID is executed by the processor once every n position interrupts. It is assumed that before execution, the motor parameters and control coefficients have already been loaded in the main processor. The algorithm calculates a current reference ($Iref$) as a linear expression of the rotational speed error (Ek), its time integral ($Integral(Ek)dt$) and its rate of change (dEk/dt).

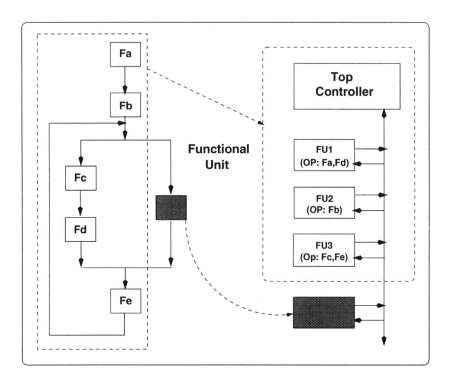

Figure 4.8 Design partitioning

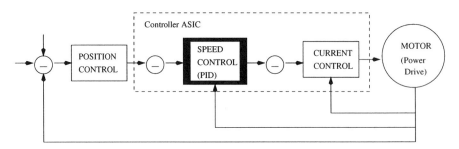

Figure 4.9 Control system with PID

The PID algorithm is given by:

$$Irefk <= (Kp * Ek) + Ki * \int (Ek)dt + Kd * dEk/dt$$

where Kp, Ki, Kd are constants and Ek is the error change. However only the close approximation given as:

$$Irefk <= (Kp * Ek) + Ki * \sum(Ek * \delta T) + Kd * \Delta(Ek)/\delta T$$

will be developed in order to be specified at the behavioral level. The integral is approximated by a sum of products while the derivative is replaced by successive position interrupts.

As introduced earlier, the goal of the system-level analysis and partitioning step is to structure the description in order to allow hierarchical description and component reuse. The result of such step is a behavioral description and its corresponding functional unit library. In fact the above PID description makes use of complex operations (*, /) that may be executed using data dependent algorithms. We will see hereafter how the components to be used in the design will be chosen and the impact of such choice on the behavioral description produced. We will even suppose for one of the solutions that the complex operators (multiplication and divisor) are not available and that they have to be designed.

The computation makes use of two complex fixed-point operations (/ and *). At this step the designer has to choose between using basic operators from the library (+, -, shift) and building specific units to perform these computations or using specific components. Three solutions will be discussed hereafter. They are represented in figure 4.10. The first solution (figure 4.10(a)) makes use of 2 components: an ALU executing +, - and a shifter executing the operations shiftleft, shiftright. The second solution (figure 4.10(b)) makes use of a multiplier, a divisor and an ALU. The third solution (figure 4.10c)) has within its set of functional modules, an ALU and a fixed-point unit (FPU) that can execute +, -, *, /.

When basic operations are selected, each instance of the complex operations has to be expanded into an algorithm that performs the corresponding computation (figure 4.10(a)). For example the multiplication procedure has to be replicated five times since the algorithm includes five instances of this operation. As stated above, such a choice may lead to a large controller in the resulting design. The other choice is to use a specific operator in the datapath that performs the multiplication. In this case each instance of the multiplication will be simply replaced by a call to this operator (figure 4.10(b)). Of course in this case we will increase the size of the resulting datapath. As the PID to be designed has no severe timing constraint, the multiplication and division operators will be implemented by sequential procedures using basic operators from the library

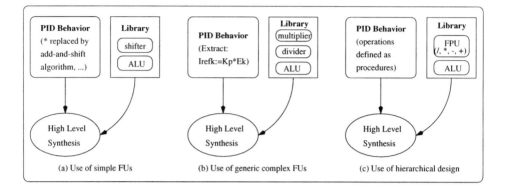

Figure 4.10 Partitioning of the initial PID description

(+, -, shift). In order to share the basic operators, the fixed-point operators (*, /, +, -) within a fixed-point unit reused in order to build the PID. This third solution is given in figure 4.10(c).

4.4 VHDL MODELING FOR REUSE

As it was stated earlier, this section will be restricted to synchronous architectures made of a top controller and a set of complex modules acting as co-processors. The top controller mainly involves the data transfers between the main controller and the components to be reused and with the external environment. This section details a VHDL writing style allowing to model this kind of architectures.

4.4.1 Modeling complex components for reuse

As explained earlier, a complex component is abstracted as a black box executing a set of operations. In VHDL these operations will correspond to procedures and functions aimed to hide the internal organization of the component. These procedures and functions will be used by the top controller as operations to communicate with the functional unit.

The abstraction of a component needs then to fix a communication protocol between the component and the top controller. This task gets complex for op-

Behavioral VHDL description styles for design reuse 143

erations whose execution time takes an unpredictable number of clock cycles. It is then necessary to define an exchange protocol made of several communication primitives. The abstraction necessary to achieve a high level view of a given module is difficult to mechanize for complex designs.

In fact a component can be abstracted in different ways according to its utilization. For instance a DSP processor can be abstracted as a coprocessor that can execute several operations: either simple algorithms such the discrete cosinus transform and its inverse and the fast Fourier transform or more complex algorithms such as an MPEG decoder which includes filtering, extraction and control functions among others. Therefore the modeling of a complex component for reuse depends on the usage made of the component.

Three main types of functional units can be distinguished:

- Simple computation functional units. The latters can execute simple computation operations with no side effect. All their operations can be written as VHDL procedures and functions returning results depending only on their parameters. Therefore they can be encapsulated within VHDL packages. A typical functional unit of this type is an ALU.

- Functional units with local memory. These correspond to units executing a set of operations that share some common data. An example of such a functional unit is given by the system defined by architecture behaviorA within figure 4.11a, where the different computations for different values of com make access to the data tmp. The abstraction of behaviorA is given by figure 4.11b.

- Functional units with local inputs and outputs. Such units allow input and output operations independent from the top controller. The system defined by behaviorB of figure 4.11c is an example of independent inputs and outputs with respect to the top controller. This means that once an operation gets to completion, its results shall be put on the output ports for recovery.

```
ENTITY system IS
PORT ( ...                    -- input values
    com    : IN type_com;     -- operation asked
    sel    : IN bit;          -- enable signal
    Z, ... : OUT type_val;    -- output values
        flag : OUT bit );     -- output validation signal
END system;
```

```
ARCHITECTURE behaviorA OF system IS
BEGIN
PROCESS
  VARIABLE tmp : type_val; -- internal global variable; result buffer
  ...
BEGIN
  WAIT UNTIL sel='1';
  CASE com IS
  WHEN com_val_op => -- operation_call
                    flag <= '0';
                    operation_call(input_values);
  WHEN com_val_rec => -- result_recovering
                    Z <= tmp;
                    flag <= '1';
  WHEN com_val_... => ...
  ...
  END CASE;
END PROCESS;
END behaviorA;
```

Figure 4.11a Modeling functional unit with local memory for reuse

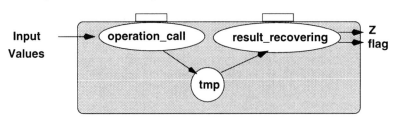

Figure 4.11b Architecture behavior A

Control signals are required for the modeling of all functional units accessed by several functions or procedures. They are present at the register transfer level and must thus be taken into account for the high level synthesis view. For simple computation functional units, they shall be ignored at the behavioral level as the different operations are completely independent ones. On the other hand for functional units with local memory or/and inputs and outputs, as different operations will share common data the different control signals have to be modeled for a structured behavioral description. It can be noticed that within figure 4.11a and figure 4.11c the signals *sel* and *com* have this function.

```
ARCHITECTURE behaviorB OF system IS
```

```
BEGIN
PROCESS
  VARIABLE tmp : type_val; -- internal global variable; result buffer
  ...
BEGIN
  WAIT UNTIL sel='1';
  CASE com IS
  WHEN com_val_op   =>   -- operation_call
                        flag <= '0';
                        operation_call(input_values);
                        Z <= tmp; flag <= '1';
  WHEN com_val_...  => ...
  ...
  END CASE;
END PROCESS;
END behaviorB;
```

Figure 4.11c Modeling functional unit with local inputs and outputs for reuse

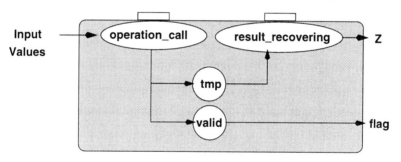

Figure 4.11d Architecture behavior B

4.4.2 Modeling for reuse with behavioral synthesis

The reuse of behavioral component will be based on the assumption that procedures and functions can be used as standard operations in the behavioral model. These are handled during scheduling and allocation like standard operations of the language (i.e. +, -, ...). In VHDL, the operations will be described as procedures and functions. For the modeling of the components themselves, several models can be proposed. We will restrict ourselves to 2 styles that can be handled by behavioral synthesis. Of course the VHDL model should also be complete in order to allow simulation at the behavioral level.

The first modeling style will be called pure behavioral VHDL model and the latter a mixed behavioral-structural model. In the pure behavioral VHDL model, a complex system is specified as a single process making use of complex procedures and functions. All procedures perform computation with no side effect. In this case the body of the functions and procedures should include the details of the computation. This model can be used only with simple computation functional units and functional units with local memory. Figure 4.12 gives the VHDL style for this model.

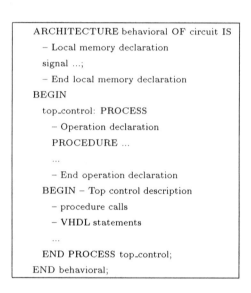

Figure 4.12 Pure VHDL modeling for reuse

The local memory of the functional unit may be declared as signals in the architecture. The use of these signals should be allowed only inside the procedure corresponding to the operations of the functional unit. This model is represented in figure 4.13 from a conceptual point of view. Each component is represented as a set of methods acting on a local data structure.

Table 4.14 gives a typical pure VHDL description. This describes the top control of the PID corresponding to solution (c). In this case, the multiplication and division are modeled as specific functional units. Since these are data dependent algorithms we use the following primitives: *start_mult*, *start_div* and *get_result*.

Behavioral VHDL description styles for design reuse 147

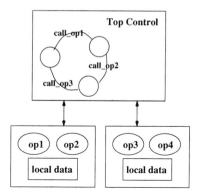

Figure 4.13 Description Representation

With this style, a component is modeled as a set of signals and a set of procedures acting on these signals. This model is close to object oriented approach where objects are data structures. In order to be used by behavioral synthesis, the synthesis tools need to provide extra capabilities allowing to handle complex operations. In addition to the behavioral description, the behavioral synthesis needs to know the correspondence between the procedure and the functional unit. This can be provided in a synthesis view as explained in section 4.2.2.

The second modeling style is called the mixed behavior-structure model. A functional unit is modeled as a component whose access is restricted to a set of procedures. This is the most general model. It allows to handle all kinds of black boxes; however it is more difficult to handle by behavioral synthesis. In the mixed model the VHDL description is made of a top controller given as a process and a set of instances that represent the components, as in figure 4.15

This model is represented in figure 4.16 from a conceptual point of view. Each component has a local control, a local I/O that can be connected to the external world or to another functional unit. The top controller can access each component through a set of predefined functions and procedures. These are used to exchange data between the top control and the components.

```
ARCHITECTURE behavioral OF pid IS
  - Declaration of memory
  ...
BEGIN
  main : PROCESS
    - Variable declaration
    VARIABLE tmp: ...;
    - Procedure declaration
    PROCEDURE start_mult(A,B: IN ...) IS
    BEGIN
      tmp:=A*B;
    END start_mult;
    PROCEDURE start_div(A,B: IN ...) IS
    BEGIN
      tmp:=1/A;
    END start_div;
    PROCEDURE get_result(A: OUT ...) IS
    BEGIN
      A:=tmp;
    END get_result;
    ...
  BEGIN - Behavioral description
    ...
    WHILE .. LOOP
      ...
    END LOOP;
    ...
  END PROCESS main;
END behavioral;
```

Figure 4.14 Pure behavioral architecture

4.5 TOWARDS OBJECT ORIENTED DESIGN IN VHDL

The two modeling styles introduced above are conceptually very close to object oriented modeling. However present VHDL does not provide a natural way for expressing this kind of modeling style. Lots of recent work are targeting OO-VHDL extensions. There are two main approaches which are in competition

Behavioral VHDL description styles for design reuse

```
ARCHITECTURE mixed OF circuit IS
    COMPONENT complex_FU PORT (...); END COMPONENT;
    ...
    SIGNAL ...;
BEGIN
    instA : complex_FU
        PORT MAP (clk, start_sig, sel_sig,
        param1, param2, param3, fin); -- concurrent statements
    main : PROCESS
    -- Procedure declaration to access FU
        ...
    BEGIN -- behavioral description
        operation_call(sig1, sig2); -- procedure calls
        WAIT UNTIL flag;
        result_recovering(sig3); ...
    END PROCESS main;
END mixed;
```

Figure 4.15 Mixed behavioral-structural architecture

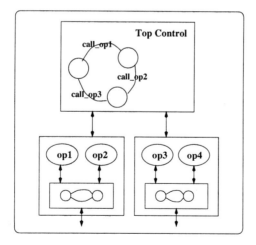

Figure 4.16 FSMC model

and that may be combined. The first [GW95] targets object orientation of data types. This model is very close to software models; the basic objects are data types. This kind of approach would make easier the modeling following the pure VHDL style introduced in section 4.4.2 (figure 4.12). The second approach supports a structural hierarchy where basic objects are components [SMG95]. This model makes easier the mixed behavior-structure model introduced in section 4.4.2 (figure 4.14).

Figure 4.17 gives a flavor of an OO-VHDL description following [SMG95]. We used the PID example. The fixed-point unit is declared as an entityobject whose architecture executes three methods. This model is very close to the abstraction model introduced. This component is instanced in the PID model as an objectcomponent. It is reused within the PID through method application (send statements). This example shows clearly that this kind of extension would make more natural modeling for reuse.

```
entityobject fpu is

    port (A,B: in ...;
          Z: out ...);

    operation start_mult (A,B: in ...);
    operation start_div  (A,B: in ...);
    operation get_result (Z: out ...);

end fpu;
--------------------------------------------
architecture oo_desc of fpu is

    instance variable tmp : ...;

    operation start_mult (A,B: in ...) is
    begin
        tmp := A*B;
    end start_mult;
    operation start_div  (A: in ...) is
    begin
        tmp := 1/A;
    end start_div;
    operation get_result (Z: out ...) is
    begin
        Z <= tmp;
    end get_result;

begin
```

Behavioral VHDL description styles for design reuse

```
end oo_desc;
-------------------------------------------
entityobject pid is

    port (...);

    ...

end pid;
-------------------------------------------
architecture oo_desc of pid is

    ObjectComponent fpu
    port (A,B: in ...;
          Z: out ...);
    end ObjectComponent;

    signal ...

begin

    object Inst_fpu: fpu
    port map (...);

    process
        variable done: bit;
        variable Irefk, Ek, Kp, ...: ...;
    begin
        ...
        send Inst_fpu start_mult (Kp,Ek);
        send Inst_fpu get_result(Irefk,done);
        while done='1' loop
            send instance get_result(Irefk,done);
            ...
        end loop;
        ...
    end process;

end oo_desc;
```

Figure 4.17 Pid description

4.6 SUMMARY

Design reuse at the behavioral level consists of the reuse of subsystems of any degree of complexity, including existing modules, RTL and logic descriptions and even bloc obtained from previous behavioral synthesis sessions. There are two ways for combining behavioral synthesis and design reuse. Existing components may be reused as external units or as coprocessors executing a set of predefined operations.

Design reuse leads to modular design and more generally to structured design methodologies which involve three main steps: partitioning the design into modules, module design and module abstraction for reuse.

Abstraction is the key concept in design reuse. Without abstraction, faced with VHDL model, designers would be trying to figure out what each component does and how to reuse it. In order to allow reuse with behavioral synthesis, the concept of behavioral component is introduced. This is an extension of the functional unit concept. In order to be reused, existing modules are abstracted as coprocessors able to execute a set of operations. Several specification views may be needed during the behavioral synthesis process.

In VHDL, the operation executed by behavioral components may correspond to standard operations (e.g. +, *, ...) or procedures and functions aimed to hide the communication with the behavioral component. Two VHDL writing styles are suggested to reuse complex components within behavioral models. The first one is called pure VHDL model; a system is specified as a single process making use of complex procedures and functions. Within this model, a component is modeled as a set of signals and a set of procedures acting on these signals. This model is close to signal based object oriented models. The second modeling style is called the mixed behavior-structure model. In this case, a functional unit is modeled as a component whose access is restricted to a set of procedures. This model is more general; it allows to handle components with a local controller and local inputs and outputs. This style is close to structured object oriented models.

5
ANATOMY OF A BEHAVIORAL SYNTHESIS SYSTEM BASED ON VHDL

This chapter details AMICAL, a VHDL behavioral synthesis tool allowing design reuse. AMICAL starts with a VHDL behavioral description and an external library of functional modules. AMICAL provides methods for abstracting large existing cores in order to reuse them as black boxes within behavioral description. It issues the FSMC model as intermediate form. This model allows to handle large functional units as co-processors. AMICAL allows interactivity in addition to a pure automatic execution mode.

The synthesis process proceeds into several refinements steps.

- Compilation of the VHDL description
- Partial scheduling and building of a behavioral FSMC
- Allocation/binding of functional units (operations and co-processors)
- Micro-scheduling and building of an RTL FSMC
- Connection Allocation
- RTL and datapath controller generation

The focus of this chapter will be more on models than on algorithms. After a presentation of the main principles of AMICAL, the metamorphosis of the design from a VHDL description into a datapath/controller architecture will be explained. The different models obtained after each refinement step will be detailed. Section 5.3 deals with interactive synthesis. This feature allows the designer to mix automatic and manual design for a larger design exploration.

154   CHAPTER 5

Finally, section 5.4 outlines key issues related to the introduction of behavioral synthesis within the design process. This includes links with the designer (debug, simulation) and links to RTL synthesis tools.

5.1 MAIN PRINCIPLES

Classical synthesis systems usually run in an almost automatic press-button manner; therefore their performances lie on the algorithms implemented. The designer has little freedom in orienting the result. This is convenient for DSP-like systems as these algorithms usually perform loop unfolding efficiently, but suits less control-flow dominated circuits. On the other hand, AMICAL allows interactively, in addition to a pure automatic execution mode; the designer can direct influence the synthesis according to specific knowledge of the circuit or to personal experiences, on mixing manual and automatic execution of the synthesis steps. Moreover existing behavioral synthesis tools usually restrict the functional unit concept to modules executing predefined operations of the initial behavioral specification, while AMICAL brings a generalization of this concept leading to the use of co-processors as functional units.

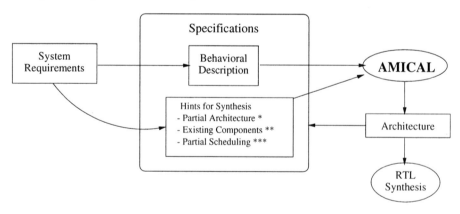

Figure 5.1 Synthesis by incremental refinements

A system design starts with a set of requirements, from which a main behavioral specification can be extracted. At this stage, the designer has often only a "rough" idea of the architecture (*) and of some of the components required. The description style accepted by AMICAL allows to take into account this kind of information in the behavioral description. This description may indicate the

Anatomy of a Behavioral Synthesis System Based On VHDL 155

use of specific components, which may correspond to existing descriptions (**) or to subsystems to be designed later. The behavioral description may also impose partial scheduling (***) or a critical part of the algorithm. Starting with an initial description, the synthesis-based design process with AMICAL is made of a refinement loop including the exploration of the possible architectural solutions. Such a process will help the designer to have a better idea of the final architecture. When a solution meeting the system requirements (and constraints) is reached, the lower level synthesis may start. Otherwise, on making use of acquired knowledge about the specification, the designer may improve the behavioral description and restart another synthesis loop. Of course this iterative scheme is only possible thanks to the high execution speed of AMICAL (high performance). In fact a typical synthesis process for a large design generally takes less than 2 minutes.

The figure 5.2 shows the incremental refinement model used by AMICAL. The initial description is a VHDL architecture with a main process. This process may call functions or procedures and may make use of a set of resources (input/output units: I/O, functional units: FU, co-processors: co-proc). The synthesis process will achieve the refinement in the controller, the datapath and the connections between them.

The design process shown in figure 5.2 illustrates two types of refinement. The first one is manual and allows the user to express architectural ideas in the input description. This kind of refinement includes the reuse of specific blocks or the partial (re)writing of the behavioral description to impose some scheduling constraints. The second type of refinement is more automatic; it allows the architectural design space exploration starting with the same behavioral description. The next section describes the synthesis steps and the intermediate design models used by the AMICAL system to go from the behavioral description to the architecture.

5.1.1 VHDL description style

Behavioral synthesis starts with a functional specification and generates an architecture composed of a data-path and a controller that may feed existing synthesis tools acting at the register transfer and logic levels. At the behavioral or architectural level, the design is specified in terms of control steps by means of programs, algorithms, flowcharts, etc. The concept of operations and control statements is used to sequence the input and output events; control statements, within VHDL description for instance, consist of instructions such as loop and

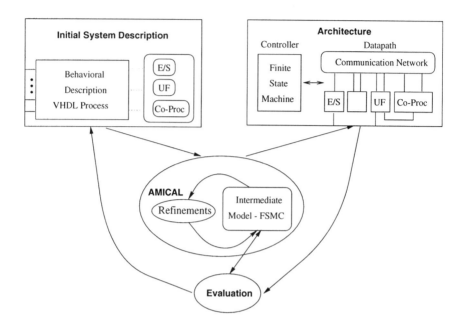

Figure 5.2 Incremental refinement model

wait statements. Typical representations at this level are behavioral finite state machines, control-flow graphs, data flow graphs and control/data flow graphs, all corresponding to event-driven specifications. Descriptions at this level are often composed of a set of protocols to exchange data with the external world and an internal computation. The computation step being itself composed of a set of operations executed between two successive input or/and output events, may take several clock cycles.

AMICAL starts with two kinds of information: a behavioral specification given in VHDL and an external functional unit library. We will give more details about the VHDL style accepted in the next section. AMICAL makes use of a behavioral synthesis model allowing to handle very large design based on hierarchical specification and multi-level scheduling. The basic idea behind this model is that a complex system is generally composed of a set of subsystems performing specific tasks. A high-level specification of such a system needs only to describe the sequencing of these tasks, consequently the coordination of the different subsystems. Each subsystem corresponding to a functional module designed (or selected) to perform a set of specific modules, is modeled as a unit executing a set of complex operations. Therefore the behavioral specification may be seen as a coordination of the activities of the different subsystems. The decomposition of a system specification into a global control and detailed tasks allows to handle very complex design through modularity.

The behavioral description of a complex design can be represented by figure 5.3. This is made of a main process, a set of component instances and a glue representing the rest of the description. More details about behavioral VHDL is given in section 5.2.1.

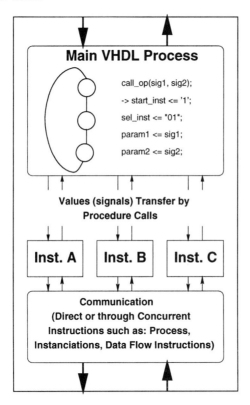

Figure 5.3 Behavioral description structure

5.1.2 Target architecture

The backbone of any automatic design is the architectural model used. AMICAL is based on the FSMC model. This is a flexible architecture model allowing modularity and design reuse. The target architecture of AMICAL is shown in figure 5.4. It is composed of a top controller, a set of functional units and a communication network. These last two constitute the datapath. The communication network is composed of buses, multiplexers and registers. The network is built in order to allow the communication between functional units, and with the external world. The number of buses and multiplexers is fixed after scheduling and allocation.

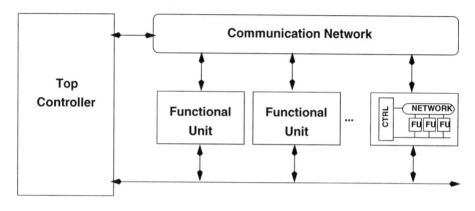

Figure 5.4 Target architecture

The top controller sequences the operations executed by the functional units and the communication network. The controller is generated automatically during the synthesis process. The architecture may include several functional units that may run in parallel. The amount of parallelism is fixed during the synthesis process. The functional units interact through a network of communication which is composed of buses and registers. FUs are handled as co-processors aimed to execute the operations of the behavioral description. With this scheme, large memorization blocks and I/O units are handled as functional units and managed by the user in the behavioral description. This architecture is general enough to represent a large class of designs. The target may be a simple ASIC or a complex application specific architecture where the top controller acts as a main processor and the functional units as co-processors. In the present version, the controller is a simple hardwired finite state machine. However, this model can be extended to handle programmable controllers. This architecture model allows both design reuse and modularity. Functional units are used as black boxes. They may correspond to already existing blocks or may be themselves the result of a previous synthesis step. The use of this target architecture avoids to restrict the functional units of the library to simple logic and arithmetic operators. In real life circuit, it is generally the case where a design includes complex operators such as cache memories, I/O units, etc. These generally correspond to existing hardware and may execute several operations.

5.1.3 Design flow

The AMICAL design flow is illustrated by figure 5.5. The two kinds of information required for synthesis are a behavioral specification and a library of functional units. AMICAL then generates a register transfer level (RTL) description that can feed existing RTL and logic synthesis and simulation tools. The behavioral description, given as a standard VHDL file, may make use of complex subsystems through procedure and function calls. However for each procedure or function used, the library must include at least one functional unit able to execute the corresponding operation. During the different steps involved in the behavioral synthesis, the functional units are used as black boxes. Although only a VHDL subset is accepted, the latter is large enough to allow large and complex design specifications. The different steps involved in the synthesis process are: scheduling, allocation and architecture generation. The process includes two allocation/scheduling steps.

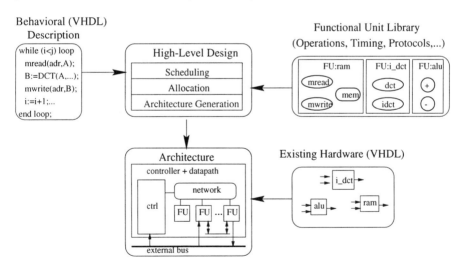

Figure 5.5 AMICAL design flow

All the synthesis steps are based on a common intermediate model called FSMC (see section 2.1.2). The first step reads in the VHDL description and produces an FSMC model that will be used by the other synthesis steps. This transformation performs a partial scheduling of the behavioral description in order to decompose the initial specification into a set of "macro-control-steps" where parallel tasks are performed. A task at this level may require several clock cycles. The FSMC is them refined through functional unit allocation, micro-scheduling, connection allocation and controller/datapath generation.

5.2 DESIGN STEPS AND EXECUTION MODELS

The synthesis process that leads to a register transfer level description from a behavioral description includes many intermediate steps. The two main transformations applied are scheduling and allocation. All these steps act on a common intermediate form called FSMC.

In this chapter we detail the different architectural models involved during the synthesis steps by the AMICAL system. The AMICAL design flow executes two interleaved scheduling and allocation steps. In other words, there is an allocation step between the two scheduling phases, and vice-versa.

Each synthesis step refines its input description by adding new details. The behavioral synthesis produces a register transfer level description that defines an architecture, starting from an algorithm as input.

As introduced in section 5.1.3 and shown in figure 5.6, the synthesis steps done by AMICAL are in the following order: macro-scheduling, functional unit allocation, micro-scheduling and the connection allocation. Each step is an independent task. They appear in the design flow as refinement steps in the process of going from the behavioral description to the register transfer level description. Table 5.1 shows the objects handled by the different architectural models used during the synthesis process. The different models and steps will be detailed in the next section.

5.2.1 Input description

The synthesis process requires only two kinds of information: the algorithm or the function to be executed and a library of functional units. These functional units can execute multi-cycle operations and can also have been generated automatically by a previous high level synthesis step.

Anatomy of a Behavioral Synthesis System Based On VHDL 161

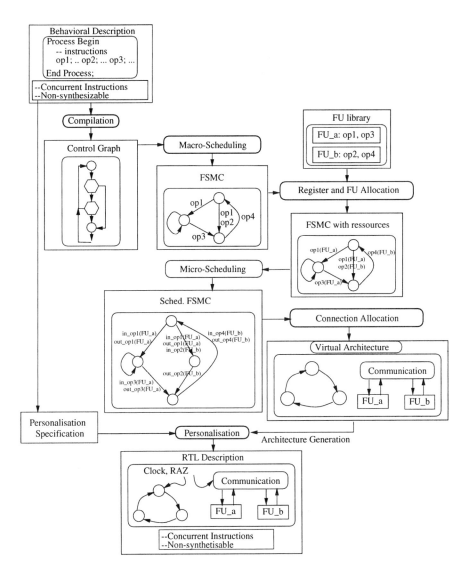

Figure 5.6 Synthesis flow

The user may write the behavioral specification a *priori* with no influence of the target architecture; he may only describe the algorithm in a sequential format. Nevertheless, each operation in the initial behavioral description may require the allocation of a functional unit, therefore the specification can directly in-

Architectural models	Organization	Time Base	Handled objects
Input Behavioral Description(VHDL)	process A given FU set	Calculation Step	Inputs/Outputs, Signals,Variables, sub-programs, Operations,...
Control Graph	Control and Data Flow Graph	Control Step	Actions, Tests, Conditions
Behavioral FSMC	Finite State Machine	Macro-cycle	States, Conditions, Actions
FSMC with Resources	Finite State Machine + Allocated Resources	Macro-cycle	States,Conditions, Actions,FUs, Registers
RTL level FSMC	Finite State Machine + Allocated Resources	Micro-cycle	States,Conditions, Transfers,FUs, Registers, Connection Elements
Virtual Architecture	Controller + Datapath	Micro-cycle	Instances, Connection,...
RT level Description	Controller + Datapath	Clock cycle	Instances, Finite State Machine

Table 5.1   Architectural models

fluence the architecture. The elaboration of the behavioral description and the functional unit library are tightly linked. We will discuss this interdependency with the aid of an application in chapter 6.

Behavioral VHDL

The behavioral synthesis done by AMICAL conceives an algorithm described by a process in a standard VHDL file. In the case where the VHDL behavioral description has more than one process: other processes and/or concurrent instructions as the instances or simply data flow instructions, AMICAL will treat only one process. Nevertheless AMICAL offers the means to do the integration of the part nonsynthesized with the resulting synthesized RTL part (see section 5.4.5).

The VHDL subset accepted by AMICAL in the behavioral input description includes a very large number of sequential instructions to allow the description of complex algorithms that will lead to corresponding large and complex circuits. In these subsets there are the conditional operations ("**if ... then ... end if;**", "**if ... then ... else ... end if;**", "**case ... when ... end case;**"), the loops ("**loop ... exit loop; ... end loop;**", "**while ... loop ... end loop;**"), the procedure and function calls, the variables and signals expressions, and the wait instructions ("**wait until ...;**", "**wait for ...;**").

A procedure may be compiled in two ways:

- expanded inline, the procedure call is replaced by the body of the procedure.
- interpreted as complex operation that has to be executed on a functional module of the data-path.

The extension of the concept of operation to procedure and function call is a major extension of the high level synthesis process. Better yet, this feature allows a flexible compilation method. In opposition to the classic synthesis tools that accept only input descriptions with basic operations like +, -, *, ..., the procedure call gives access to an infinite number of complex operations.

The behavioral description accepted by AMICAL may include several wait instructions inside the same process. It is possible to combine many wait statements on different signals and complex conditions with loops and complex control instructions.

The use of multiple wait statements inside the same process brings a great benefit. It allows an easy description of algorithms with complex data exchange protocols with the external world (through the pins, or more locally, through signal sharing between operations concurrent with the process). Figures 5.7(a) and 5.7(b) show two examples that correspond respectively to an example of a typical behavioral description and an excerpt of the VHDL code of an answering machine. The description in the figure 5.7(a) has three distinct blocks. First, a calculation loop that contains a wait instruction found at each interaction. It is preceded by the initialization and reading of the inputs and followed by a synchronization step that sends to the external world the result of the calculation. This kind of behavioral description is frequently used. The register transfer level description of such a behavior would need a much larger VHDL description. The different wait statements have to be transformed into a combination of loops and waits on the clock signal. The combination of waits and loops have to be removed by a scheduling transformation The algorithm shown in the figure 5.7(b) is an excerpt of an actual design case: an answering machine which will be used through this section to illustrate the different execution models found during the synthesis. It has a combination of control loops, waits and loop exits. This is the kind of combination that is normally forbidden in RTL descriptions.

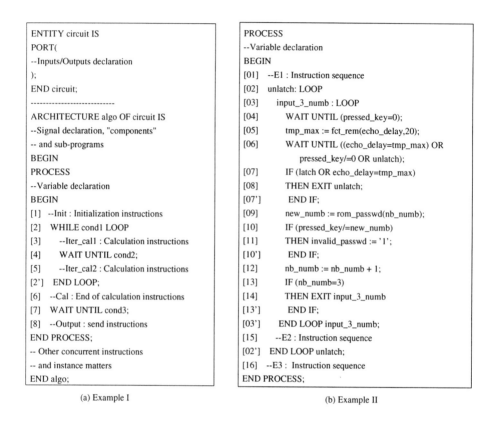

Figure 5.7 Examples of behavioral descriptions

Library of components

AMICAL is able to use two types of components:

- a set of basic components. These elements correspond to the basic architecture elements, as the registers, the multiplexers, ...

- an extendible functional unit library.

AMICAL uses a library of basic components. The elements of this type have their functionality fixed; the system knows the different connection elements that could be instanced. Their function and interface protocol are all predefined. Each one of these components has a generic bit width. The list of these elements (shown in figure 5.8) is enumerated here:

- Register with one output that connects to other component in the datapath (VariableRegister)

- Register with two outputs, one like the previous register and the other one for connection to the controller (StatusReg)

- Constant of generic value (ConstReg)

- Interface port between the external world and the datapath, or vice versa (ExtCon_IN, ExtCon_OUT); this interface could be a simple connection or also a complex function like signal amplification, coding or decoding

- Multiplexer with N inputs (where N; the number of inputs is defined by AMICAL to get the optimal one: Mux_2/ Mux_3/ .../ Mux_i/ .../ Mux_N)

- Tri-state gate, essential to the bus based architecture

- Interface cell between the controller and the datapath, or vice versa. The int_cell have memory elements depending on the anticipation required in the generation of the control signals to the datapath.

Technology modeling

The technology modeling description file contains the size of the components, the performance parameters (maximum consumption and maximum delay), and the constraints. They will be useful to estimate the performance of the synthesized structure and to evaluate its quality. Figure 5.9 gives an example of the technology modeling file.

5.2.2 Library of functional units: behavioral components

In order to allow reuse, we need to abstract complex components. The concept of behavioral component is a generalization of the functional unit concept. A behavioral component allows the use of existing macro-blocks in the behavioral specification. A functional unit may execute standard operations or new customized operations introduced by the user. A functional unit can be called

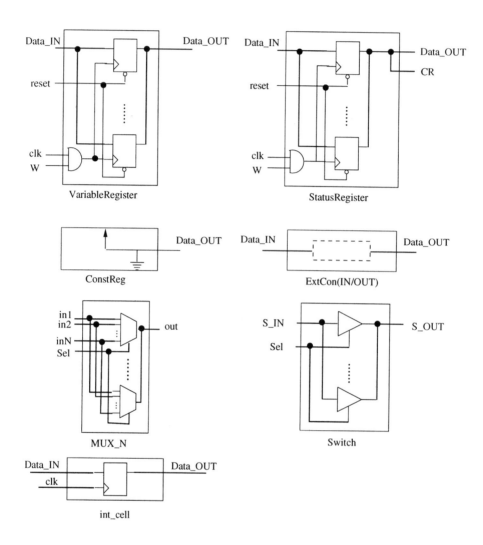

Figure 5.8 The base components on the library

within a behavioral description through procedure and function calls. It can accept and return parameters.

Each functional unit can be specified at four different abstraction levels: the conceptual view, the behavioral view, the implementation view and the high-

Anatomy of a Behavioral Synthesis System Based On VHDL   167

```
 1   ( TECHNOLOGY_FILE
 2     ( PARAMETER
 3        ( CONSTANT_REGISTER (WIDTH 30) (HEIGHT 60.0)
              (AREA 20.0) (POWER 15 15) (MAX_DELAY 1))
 4        ( FLAG_REGISTER (WIDTH 640) (HEIGHT 60.0)
              (AREA 240.0) (POWER 55 5) (MAX_DELAY 4))
 5        ( VARIABLE_REGISTER (WIDTH 640) (HEIGHT 60.0)
              (AREA 240.0) (POWER 55 5) (MAX_DELAY 4))
 6        ( EXTERNAL_REGISTER (WIDTH 640) (HEIGHT 60.0)
              (AREA 240.0) (POWER 0 0) (MAX_DELAY 2))
 7        ( SWITCH (WIDTH 320) (HEIGHT 60.0)
              (AREA 110.0) (POWER 15 5) (MAX_DELAY 2))
 8        ( MUX (WIDTH 200) (HEIGHT 60.0)
              (AREA 88.0) (POWER 20 5) (MAX_DELAY 3))
 9        ( BUS ( HEIGHT 10.0) ( POWER 10 10)( MAX_DELAY 1))
10     )
11     ( CONSTRAINT
12        ( MAX_MUX 50 ( WEIGHT 1))
13        ( MAX_MICRO 100 ( WEIGHT 1))
14        ( MAX_BUS_CHANNEL 9 ( WEIGHT 10))
15        ( MAX_FU 10 ( WEIGHT 5))
16        ( MAX_SWITCH 90 ( WEIGHT 5))
17        ( MAX_WIDTH 500.0 ( WEIGHT 0))
18        ( MAX_HEIGHT 100.0 ( WEIGHT 0))
19        ( MAX_AREA 50000.0 ( WEIGHT 10))
20        ( MAX_POWER 2000 ( WEIGHT 5))
21     )
22     ( FACTOR
23        ( FDATAPATH 1.2)
24        ( FCONTROLLER 0.44)
25        ( FCIRCUIT 1.3)
26        ( TR2AREA 0.0002)
27        ( SWPOWER 1.3)
28     )
29   )
```

Figure 5.9   AMICAL technology file example

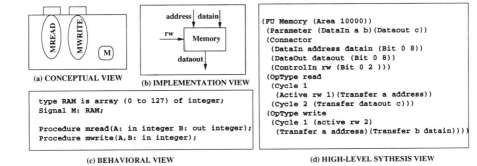

Figure 5.10   The four abstraction levels

level synthesis view. Figure 5.10 shows these four views for a memory cell that can achieve the 2 operations: mread and mwrite.

From the conceptual point of view, the functional unit is an object that can execute one or several operations which may share some data (M). At the behavioral level, the functional unit is described through the operation that can be called from the behavioral description. These may correspond to standard operations or procedures and functions. The behavioral view in figure 5.10 is a VHDL package that includes the procedures mread and mwrite. The implementation shows an external view of a possible realization of the functional unit; it thus includes the different connections of the functional unit: inputs, outputs and selection commands (selecting the procedure to execute).

The high-level synthesis view of the functional unit links the behavioral and implementation views. It includes the interface of the functional unit, its call-parameters (corresponding to the operation parameters), the operation set executed by the functional unit as well as the parameter passing protocol for each operation. This protocol is expressed through static clock cycles, each operation needs to have a fixed predictable execution time. In order to overcome this constraint and to enable the use of complex functional units that may execute operations with data-dependent execution time, the methodology used consists in splitting the operation in a set of atomic operations with fixed execution time. The behavioral description will then be written according to the atomic operations introduced. During synthesis, the designer may need to ensure the coherence of the synthesis results through hints within the specifications.

The designer invokes a functional unit, from the library contained in his synthesis environment, through a simple procedure call. Functional units can be of any degree of complexity and can themselves be the result of a synthesis process.

Memory with complex addressing function and specific embedded computation and control can be easily described with this scheme. The addressing functions may be realized by an independent functional unit or may be integrated within the memory unit. In the same way complex I/O units may be used. They are also accessed through function and procedure calls. These may execute complex protocol or data conversion.

According to the characteristics of the application, a set of functional units will be provided before starting the synthesis process. For instance, the use of special purpose hardware units able to increase the computing power of the architecture may be provided. The correspondence between the operations of

Anatomy of a Behavioral Synthesis System Based On VHDL

the behavioral description (standard operators such as + and -, and procedure and function calls) and the functional units is made during the synthesis process. Initially we may have a library where each operation may be executed on one or more functional units and where each functional unit may execute several operations. We may even have operations that have different execution schemes on different functional units. The number of functional units selected (allocated or instanced) will depend on the parallelism required by the initial description. Of course the synthesis process tries to share as much as possible the use of the functional units.

5.2.3 VHDL compilation

The synthesis process consists of a sequence of steps that refine in an incremental way the behavioral description to produce the desired architecture. The synthesis model adopted by AMICAL stands on the fact that the designer has partial knowledge of the architecture he wants to synthesize. This information refers to the pre-defined components or to the data organization. The behavioral description could take them into account. The initial behavioral description can be seen as an abstract architecture representation, from which one can distinguish certain organization elements such as the controller and the datapath.

Some procedures and the functions shall define the access to the functional units in the VHDL process being synthesized. They imply the instantiation of functional units. The sequencing and the test conditions are parts of the controller. Figure 5.11 shows a datapath controller representation of a VHDL description. This representation will be used to illustrate the refinement of the behavioral description. This model is a symbolic representation of the FSMC model. In the example shown in figure 5.11, the procedure call_opA gets in place two control signals (that are, start_instA and sel_instA) and one data signal (param1_instA) of an instance defined in the behavioral level. This procedure call infers a functional unit (inst_FUi) that can execute the operation. This is true for all the procedure or function calls and all the utilization of standard operators.

Any behavioral description input language must first be compiled to extract the information actually needed for the synthesis. With this objective AMICAL uses the database of existing VHDL compilers. The synthesis flow includes first one step for the compilation of the behavioral VHDL. The compilation

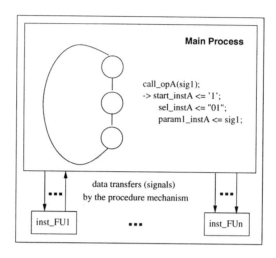

Figure 5.11 Organization of the behavioral description

step does the syntax verification of the VHDL description and produces the corresponding parse tree.

The generation of the AMICAL intermediate form, FSMC, is performed in 2 steps. First a control flow graph is generated. This will be used by the scheduling algorithm in order to produce the FSMC.

The figure 5.12 shows the control graphs that correspond to the VHDL descriptions ((a) and (b) respectively) presented in the figure 5.7. The node numbers (inside the square brackets) in the figure 5.12 correspond to the line numbers in the figure 5.7.

In the control graph, the statement sequence determines the sequencing of the operations, without explicit notion of time. This description will be used for the scheduling step and it will be transformed in a behavioral FSMC.

5.2.4 Scheduling and behavioral FSMC

The next step in the synthesis process is called macro-scheduling. This step performs a partial scheduling of the CFG and produces a behavioral FSMC.

Anatomy of a Behavioral Synthesis System Based On VHDL 171

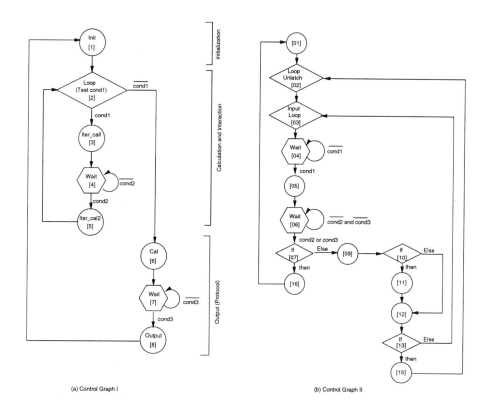

Figure 5.12 Control flow graph models

The behavioral FSMC is represented as a transition table. The operations may be of any complexity and may require many clock cycles for their execution. The operations associated with each transition of this description will be executed in parallel, however they may not require the same number of clock cycles for execution. For each transition of the behavioral FSMC, the slowest operation will define its effective delay (the number of cycles needed to its actual execution).

Each controller transition is defined by the current state, the condition to be satisfied and a set of operations or actions to be executed. In other words, many transitions could start from the current state. The condition presented will determine the transition to be done. The scheduling is a refinement of the initial VHDL model. Figure 5.13 shows an excerpt of the behavioral FSMC corresponding to the VHDL description given in figure 5.7(b).

At this moment, the datapath/control structure becomes more evident. New sequence details appear at the controller level. The calculation steps of the initial description are split in control steps that do not have either loops or data dependencies. The scheduling phase fixes also the number and the kind of the registers needed. The datapath may also contains some functional units. These correspond to units which are explicitly inferred by the behavioral description (e.g Memories) or by the user (explicit binding of procedures and function calls to specific co-processors). In this case the array rom_passwd generates a functional unit of type memory. Also new links are created between the datapath and the controller. The variables and signals of the behavioral description are bound to the registers and certain procedure calls to corresponding functional units.

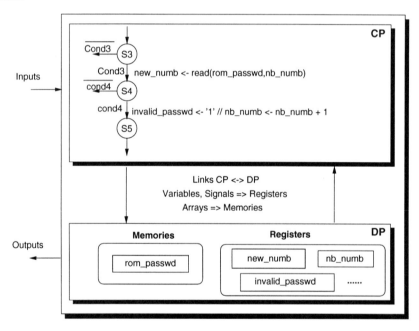

Figure 5.13 Development state after the scheduling: Behavioral FSMC

The scheduling at this level defines the role of the datapath and the controller. The treatment of the current and next states and the condition computation are done in the controller part, and the execution of operations is done in the datapath. On these operations, the scheduling fixes their execution order; the ones that could be executed in parallel will be associated to the same transition. In this case, the operations will necessarily need distinct functional units for their execution.

Anatomy of a Behavioral Synthesis System Based On VHDL 173

The controllers of the FSMC corresponding to figure 5.7 are described in the figure 5.14. Two representations are used, state tables and automaton.

State	Conditions	Actions	State Next
S1	cond1	Init; Iter_call	S2
S1	$\overline{cond1}$	Init; Cal	S3
S2	cond2	-	S4
S2	$\overline{cond2}$	-	S2
S3	cond3	-	S5
S3	$\overline{cond3}$	-	S3
S4	cond1	Iter_cal2; Iter_cal	S2
S4	$\overline{cond1}$	Iter_cal2; Cal	S3
S5	cond1	Output;Init;Cal	S2
S5	$\overline{cond1}$	Output;Init;Cal	S3

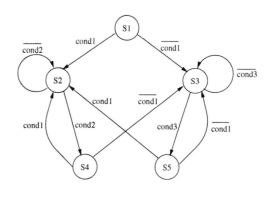

(a)

State	Conditions		Actions	State Next
S1	true		[01]	S2
S2	cond1	pressed_key = 0	[05]	S3
S2	$\overline{cond1}$	pressed_key = 0	-	S2
S3	cond2 echo_delay = tmp_max or unlatch		[16];[01]	S2
S3	cond1	pressed_key = 0	[09]	S4
S3	$\overline{cond2}$ and echo_delay = tmp_max and $\overline{cond3}$ pressed_key = 0 and not(unlatch)		-	S3
S4	cond4	pressed_key = new_numb	[11];[12]	S5
S4	$\overline{cond4}$	pressed_key = new_numb	[12]	S5
S5	cond5	nb_numb = 3	[15]	S2
S5	$\overline{cond5}$	nb_numb = 3	-	S2

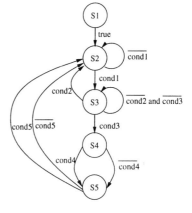

(b)

Figure 5.14 Behavioral finite state machine models

The scheduling algorithm used is based on DLS (see section 3.3.2). The algorithms explore execution paths of the VHDL description in order to find the cluster of operations that may be executed in parallel. Two operations can be executed in parallel if and only if:

- They belong to the same execution path

- There is no blocking data dependency between the first operation and all the operations on the path leading to the second operation
- there is no wait statement on the path between the two operations

The details of the algorithm are given in section 3.3.2.

5.2.5 Functional unit allocation

The scheduling of the initial description is followed by the allocation and binding of the resources (or functional units) that will ensure the execution of the operations. This step starts with a behavioral FSMC and produces a behavioral FSMC with resources (figure 5.6).

The functional unit allocation decides on the number and types of functional units needed to execute the operations of the behavioral description according to the scheduling. This step performs also binding. It associates to each operation present at the behavioral level a functional unit. In the case in which two operations must be executed in parallel, they will have two different functional units associated to them. Operations that belong to different control steps can share the same resources. Figure 5.15 shows an example of link between operations and resources. It is based on the behavioral example of the figure 5.7(b), which corresponds to an extract of an answering machine. Procedure calls are associated to co-processor able to execute them (fct_rem is associated with COMPUTE_REM). Array accesses are associated to memories. Operations (+,-) are associated to operators and ALUs. All these resources, co-processors, memories and operators and ALUs came from the library of functional modules.

This allocation/binding step produces a "behavioral FSMC with resources". Each operation of this FSMC has an associated functional unit to execute it. These functional units built the set of allocated functional units. They correspond to the components in the final architecture.

This refinement step is done entirely within the datapath. There is no supplementary information added to the controller (figure 5.16). In this example this step allocates an extra operator (adder) in order to execute the '+' operation.

At this point, the datapath is enhanced with new allocated functional units. The links between the controller and the datapath are enhanced by the correspondence between the operations and the operators.

Anatomy of a Behavioral Synthesis System Based On VHDL 175

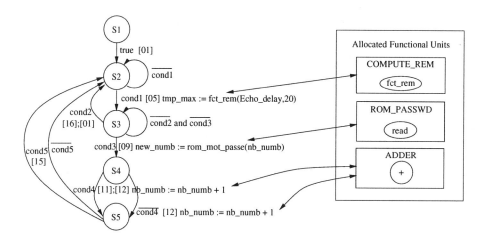

Figure 5.15 Example of link between operations and resources

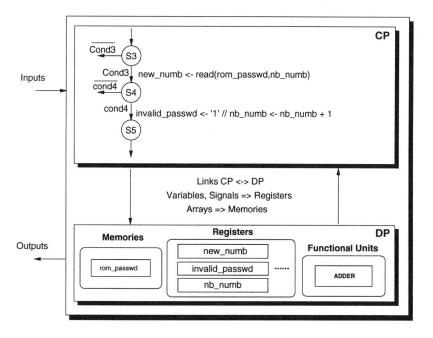

Figure 5.16 Behavioral finite state machine with resources model

5.2.6 Micro-scheduling and RTL FSM

The micro-scheduling allows the translation of a behavioral FSMC with resources to a RTL FSMC. The RTL FSMC is also represented as a transition table. However, in this case, each transition requires exactly one clock cycle to execute. Then all complex operations need to be split into single-cycle operations. The allocation/binding of functional units to operations allow to know the execution model of each operation at the register transfer level. This execution model is given by the abstraction of the components as it was introduced in section 5.2.1. In the behavioral FSMC, transitions may include complex operations that require more than one cycle to execute. Micro-scheduling allow to split these transitions into basic transitions that can be executed in a basic clock cycle. This step refines the behavioral FSMC by introducing extra cycles and states according to the execution scheme of the operations.

In the answering machine case, if we suppose that with the allocated adder, the addition requires two clock cycles, then for each transition that includes an addition (for example, the transition S4->S5) there will be a new state inserted (figure 5.17). In the same way, if the function fct_rem needs three clock cycles, then we need to create and insert two extra states within the transition S2->S3.

This step refines the controller only. The controller is now described at the clock cycle level. Each transition would need a single clock cycle to execute. At this level each transition is made of a set of parallel transfers (figure 5.18). Of course parallel datapaths are needed to allow the parallel execution of these transfers.

5.2.7 Connection allocation

The objective of this step is to generate the connection paths (or datapaths) required for the execution of all the transfers. A connection path consist of a set of interconnected elements that permits a transfer. In the AMICAL target architecture, two connection structures are available: a bus-based and a multiplexer-based. For both cases, the connection paths that correspond to parallel transfers must be independent to be consistent with the behavioral description.

Both styles allow the same instruction set, i.e. the execution of the same kind of transfer. The set of allowed transfers may be generated from the graph of

Anatomy of a Behavioral Synthesis System Based On VHDL

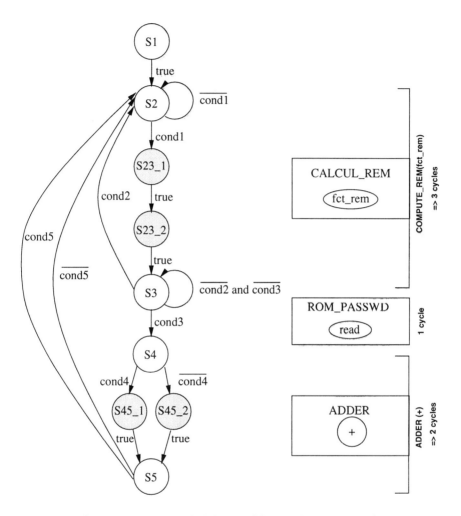

Figure 5.17 Micro-scheduling model for multi-cycle operations

figure 5.19 (see chapter 2 section 2.3). A transfer is composed of a source and destination. Each arc of figure 5.19 corresponds to a type of transfer.

This model forbids direct transfers between two functional units and between two external units (without going to an intermediate register). However it allows direct transfers between two registers.

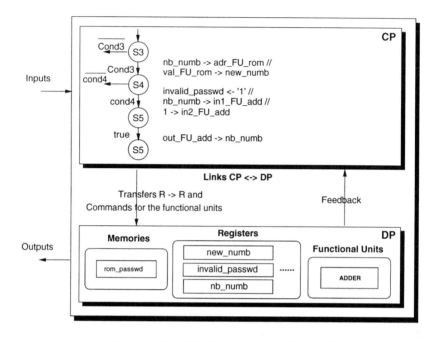

Figure 5.18 RTL finite state machine model

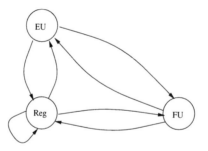

Figure 5.19 Authorized transfers

Connection allocation for bus based architecture

In the bus based architecture a connection path is made of buses and switches. Switches are needed when a destination has more than one source. Each switch will need an extra control signal in the controller. The connection allocation step tries to minimize the connection costs by looking for a solution with the minimum number of switches. The algorithm consists of two sequential tasks.

Anatomy of a Behavioral Synthesis System Based On VHDL   179

The first one is to find the number of buses needed. The second one does the switch allocation to connect the buses with the components of the datapath.

The figure 5.20 shows the datapath obtained for the GCD example. It has sixteen switches and three buses.

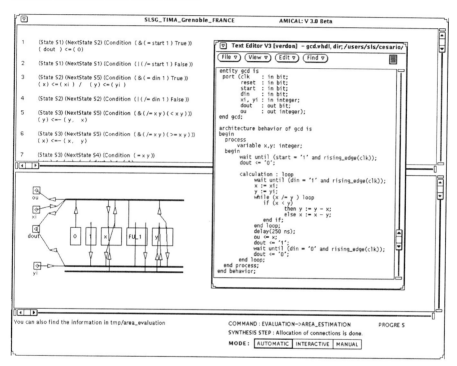

Figure 5.20   Bus-based solution for the GCD example

Connection allocation for multiplexer based architecture

In the mux based architecture a connection or path is made of multiplexers [LCGM93]. While searching the list of data transfers, the multiplexers are allocated according to the destinations of each transfer. For each destination that have more than one source a multiplexer is created. A multiplexer with N inputs will require (log2(N)) control signals in the controller.

The figure 5.21 shows the datapath obtained for the GCD example.

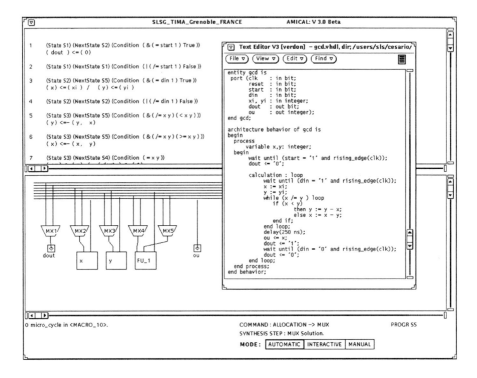

Figure 5.21 Mux-based solution for the GCD example

According to the application, one of these solutions has to be selected. The mux-based style induce less control signals in the controller and smaller surface areas in some cases. The bus based solution produces more regular datapath easier to handle by datapath compilers (easier to route).

When the input of an element of the datapath has N data sources, a Nx1 multiplexer could substitute N switches. Considering the area optimization, a multiplexer 2x1 is close to the size of a switch. Then, when N tends to a large value the mux based solution is more attractive.

In the mux based solution, the communication network contains only multiplexers instead of buses and switches. The area of a Nx1 multiplexer is smaller than the area of the corresponding N switches. Also we need less control wire: ($\log 2(N)$ instead of N).

5.2.8 RTL datapath and controller generation

At this level, the datapath contains all the elements it needs to work well. Besides the functional units and registers already allocated, there is also the interconnection elements that allow to execute the parallel transfers without conflicts. The next step is to generate the final controller structure. This step refines each transfer into a set of control signal assignments to control the datapath.

The resulting controller contains supplementary information in comparison to the register transfer level finite state machine. Figure 5.22 shows the final structure of the datapath controller model. For instance the two transfers of transition S1->S2 in figure 5.20(a) will be translated, in case of mux-based architecture, into a set of commands:

$$CTRL\_W\_x <= \text{'1'};$$
$$x\_2\_x\_Sel <= \text{'0'};$$
$$CTRL\_Sel\_xi <= \text{'1'};$$
$$CTRL\_W\_y <= \text{'1'};$$
$$x\_3\_y\_Sel <= \text{'0'};$$
$$CTRL\_Sel\_yi <= \text{'1'};$$

The set of three first commands open a path from xi to x and set the mode write for register x. The three last commands correspond in the same way to transfer "y <= yi".

5.2.9 Programmable architecture generation

Another synthesis option allows to generate a re-programmable controller without changing the datapath architecture. This architecture has advantages over the classic controller style [RBL+96]:

- the flexibility: programming allows easy error correction;
- the re-usability: the circuit could adapt to new functions only by changing the program;

The architecture of the controller has the following elements (see figure 5.23):

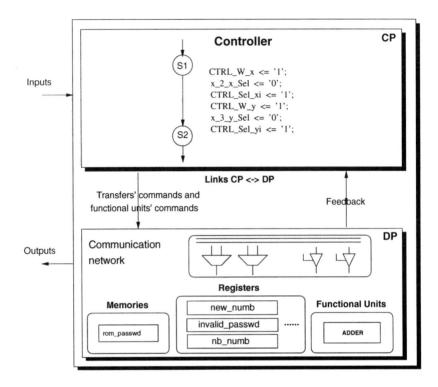

Figure 5.22 Virtual architecture

- one ROM that holds the code that will command and synchronize the datapath;

- one instruction register that holds the signals commanding the transfers in the datapath, the address of the next instruction and a indicator of the calculation mode for the next address;

- one sequencer that calculates the effective address of the next instruction as a function of the value of the instruction register.

The only synthesis steps that will change are:

- The macro-scheduling step is made with an option to execute all the conditions in the datapath (see section 3.3.3).

Anatomy of a Behavioral Synthesis System Based On VHDL

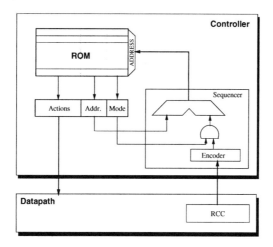

Figure 5.23 Re-programmable controller architecture

- The controller generation step uses a more sophisticated algorithm in order to split the sequencing and the operations.

This model generally induces a larger controller and datapath.

5.3 INTERACTIVE SYNTHESIS

CAD tools are useful as they enable the automation of repetitive tasks. However today designs are so large and complex that in order to speed the synthesis process, the heuristics or algorithms used are non-exhaustive and do not check for all possible combinations or solutions. Nevertheless high quality designs can still be reached in exploiting the designer's experience and knowledge. This can be achieved through interactive synthesis.

As stated in section 5.2, within AMICAL, scheduling and allocation have been split into simpler sub-tasks, enabling the designer to follow the design flow much more easily. The result of each of these sub-tasks may be easily predicted. Moreover, partial evaluations of the architecture can be done at any moment. Therefore their implementation within the AMICAL system makes the latter an interactive tool and acts as an assistant for design towards the user.

Besides the automatic synthesis mode, some of the synthesis steps may be performed manually by the user. This is done under the control of the system in order to detect inconsistent operations.

The combination of automatic and manual synthesis allows a quick and broad exploration of the design space in real time. The response time of AMICAL is very short, making it a genuine interactive system.

AMICAL is organized as an architectural synthesis environment. It works as a design assistant, combining automatic, manual and interactive synthesis. The interaction with the system is performed through a mouse and a graphical interface.

AMICAL's user-interface consists of 3 windows:

- Control window (Top): This is generally used to show and edit the controller.

- Data-path window (Middle) : It is used to show and edit the data-path.

- Information window (Bottom): It is used to print information and error messages. It also provides information about the progress of the synthesis process (last command, synthesis step, contents of the other windows, etc.). This information is needed in order to help the user during long synthesis sessions. This window also shows the synthesis mode (Automatic, Interactive or Step-by-step, Manual).

AMICAL allows to maintain the coherence between the information included in these three windows, i.e. the different aspects of the design.

Figure 5.24 shows a screen dump that gives a flavor of AMICAL at work. The right-hand window shows a VHDL description of the algorithm being synthesized. The top window contains the behavioral FSMC generated by the scheduler. The middle window shows the data-path, as synthesized by AMICAL. The bottom window provides information on the current status of AMICAL. AMICAL makes use of concepts similar to those used in CORAL II [TDW+88] in order to link the behavior and structure.

In this way, AMICAL is able to maintain the coherence between the information contained in the two top windows. For example, the designer may require information concerning the correspondence between the controller and the data-path. In figure 5.24, information about the resources used for the execution of the two parallel operations in transition 7 of the controller representation have

Anatomy of a Behavioral Synthesis System Based On VHDL   185

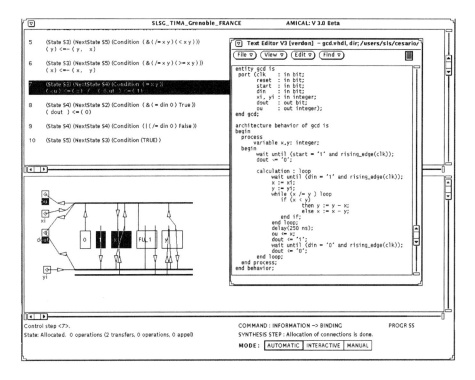

Figure 5.24   AMICAL at work

been asked (highlighted in top window). AMICAL has highlighted the appropriate data-path components in the middle window. The resources highlighted by AMICAL not only include the registers and functional units necessary for storing the variables and executing the operations, but also the buses, switches or multiplexers used in transferring the data values to and from the functional units. More information is given about the control step in the bottom window. The interaction with the user is similar to the MIES system [NSM89].

5.3.1   Mixing manual and automatic synthesis

For some of the synthesis steps executed by AMICAL, several options are offered to the designer. Starting from a behavioral-level VHDL description, the designer has the option of synthesizing entirely automatically, manually or combining automatic and manual steps in an interactive mode. This philosophy of allowing the designer to influence the synthesis process as much as he desires, makes the synthesis steps easy to understand.

AMICAL provides an initial schedule and then allows the designer the freedom to mix manual and automatic modes in order to complete the design. This section describes the facilities provided by AMICAL in order to ease the mixing of manual and automatic design.

Mixing manual and automatic design can be performed through two modes: a true interactive mode and a manual modification of automatic synthesis mode. All allocation algorithms are iteratively constructive, allowing the user to intervene at each iteration, both to modify the results or to cancel them completely. The scheduling tasks (scheduling and micro-scheduling) are performed automatically. The designer can manually modify the results of the schedule.

Figure 5.25 shows the design flow for a typical interactive synthesis step. The designer can choose to perform the synthesis task entirely automatically, step-by-step or manually. If the first option is chosen, AMICAL performs the requested task in one shot. If either of the other two options is selected, the user can perform the synthesis task manually through AMICAL graphical interface, or AMICAL can automatically perform a single iteration of the appropriate algorithm.

After each pass, the designer can examine the results and decide whether to finish the task automatically or to continue interactively. In the case of manual design, the system will verify all modifications and only correct decisions will be accepted. The interactive mode may be used when the synthesis task is performed by an iteratively constructive algorithm. Such an algorithm starts with an initial solution (that may be empty) and performs each task step-by-step. At each step, a new element (component, binding, bus, connection path, etc.) is added to the initial solution.

This scheme allows the designer to alternate between automatic and manual design throughout the synthesis process. The designer can automatically (or manually) complete a partial design started manually (or automatically). The designer may start a design manually and ask AMICAL to finish it (switch to the automatic mode). On the other hand, he may start the synthesis in the step-by- step mode and then finish it manually.

AMICAL allows the designer to mix manual and automatic design in order to produce an efficient synthesized design. Scheduling algorithms are executed completely before modifications can be made. There are essentially two schedul-

Anatomy of a Behavioral Synthesis System Based On VHDL

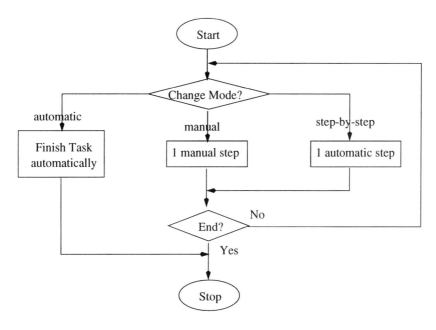

Figure 5.25 Design flow for interactive synthesis

ing tasks in AMICAL. The first is called macro- scheduling. It uses Dynamic Loop Scheduling to generate a FSMC from a behavioral-level VHDL description. This FSMC may be modified through manual interventions in order to optimize the use of available clock cycles. The second is called micro-scheduling. It allows the design to be refined through the re-ordering of transfers (micro-scheduling). The allocation tasks allow the designer to tackle the functional unit allocation, the connection allocation and the placement of components, both independently and interactively. All the allocation algorithms are based on iteratively constructive approaches, thereby allowing true interactive synthesis for mixed manual/automatic design.

A useful feature of AMICAL is the ability to manually re-schedule the design. By default, the operations in each state generated by the scheduler are executed in parallel, using different functional units. The slowest operation will determine the overall execution time of the transition. This factor implies that there may be several functional units that are idle for a lot of time, as their operations execute in less cycles than needed by the transition. By manually serializing such operations, the designer may be able to save the use of FUs. This may or may not be at the expense of the overall execution time of the transition. For example, if a transition contained a multiplication operation taking 4 cycles and two additions, each taking 2 cycles, the latter two operations may share the same adder without any loss of performance.

During each synthesis step, whatever the execution mode, AMICAL ensures the functional coherence of the circuit treated. Consequently any manual (or interactive) user's intervention gives rise to an automatic verification by the AMICAL system. For instance, during manual re-scheduling, only modifications leading to no data dependency violation are accepted end executed. During a manual functional unit allocation, functional unit binding is executed only once the system has checked that the functional unit chosen can execute the given operation.

The figure 5.26 shows an AMICAL screen copy during a synthesis session. The functional unit allocation is being executed in the manual mode. Due to the interactivity offered to the designer, the latter can get knowledge of the operations (shown within the upper window) that can be executed by the functional unit selected from the available library (through the middle window). In this figure, the functional unit chosen is called AS and as indicated within the lowest window, AS has a cost of 434 transistors; this corresponds to its size. Moreover this window shows that the subtraction is the only operation that the functional unit can execute in the given synthesized circuit. Among all the operations that can be executed by the functional unit, only the ones included in the set of operations called by the synthesized description are shown. For instance the AS (adder-subtractor) can also execute addition but this latter operation in not shown.

5.3.2 Architecture exploration

The size of a circuit description increases when its abstraction level is decreased. For example, when we shift from the behavioral level to the register transfer level, this increase may reach up to a ratio of 10 or even 20. The simulation and synthesis time being proportional to the size and complexity of descriptions, the cost of an architectural exploration at a low level is getting high and therefore it becomes necessary to reduce the number of iterations at the logic level or lower ones.

The interactive execution of the synthesis steps in AMICAL allows a wide architectural exploration for a given behavioral description and in a short time.

Anatomy of a Behavioral Synthesis System Based On VHDL

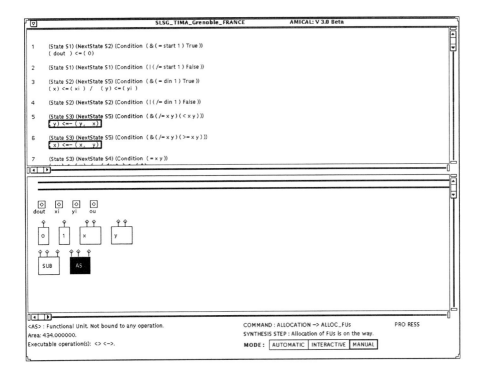

Figure 5.26 AMICAL at work

The solutions proposed can be evaluated and compared in order to define the most suitable architecture.

AMICAL allows to evaluate rapidly the relative costs of each architecture. It provides several evaluation functions aimed at on-line evaluation of the synthesis tasks already executed. This evaluation summarizes the hardware allocated and verifies that certain constraints have been met.

Statistics about the use of resources are also provided. These are the results of static analysis. The statistics file provides the number of cycles where each resource is used. For complex designs, these files provide useful indications that may guide the designer's decisions. For example, if the number of buses allocated exceeded the imposed maximum, in order to reduce the number of buses, we might perform some manual micro-scheduling to remove some of the parallel transfers.

Figure 5.27 shows an evaluation and a statistic report for the architecture given in figure 5.21.

```
(Evaluation of the synthesized data path
  (Filename of macro-cycle description : gcd)
  (Filename of FUs library : gcd)
  (Allocated registers (number : 4) (NBTR : 480)
    (Variable register :)
    (Constant register : <0> <1>)
    (Flag register : <x> <y>)
    (MAX_NUMBER 10) (WEIGHT 1)
    (ESTIMATION on register allocation : SUCCESS)
  )
  (Allocated FUs (number : 1) (NBTR 640)
    (Allocated FU <FU_1> == <SUB> (NBTR : 640))
    (MAX_NUMBER 4) (WEIGHT 1)
    (ESTIMATION on FU allocation : SUCCESS)
  )
  (Allocated connections (NBTR 528)
    (Allocated muxes (number : 5)
      (MAX_NUMBER 10) (WEIGHT 1)
      (ESTIMATION on multiplexer allocation : SUCCESS)
    )
  )
)
(Evaluation of the synthesized controller
  (Number of transistions 10)
  (Number of states 5)
  (Number of input-outputs 47 (unit bit))
  (FACTOR FCONTROLLER 0.44)
  (Controller NBTR: 341 transistors)
)
(Total area (NBTR : 1989 transistors)
      (The factor TR2AREA : 0.000200)
      (area : 0.40 mm2)
  (MAX_AREA 3000.00) (WEIGHT 10)
  (ESTIMATION on total area: SUCCESS)
)
)
(FINAL_ESTIMATION : SUCCESS)
)
```

(a) evaluation report

```
(Statistics of the synthesized data path
  (Filename of macro-cycle description : gcd)
  (Filename of FUs library : gcd)
  (Total number of macro-cycles : 10)
  (Total number of micro-cycles : 6)
  (Variable Registers (Total number : 0)
  )
  (Constant Registers (Total number : 2)
    (Name 0 (Reading : 2))
    (Name 1 (Reading : 1))
  )
  (Flag Registers (Total number : 2)
    (Name x (Reading : 3) (Writing : 2))
    (Name y (Reading : 2) (Writing : 2))
  )
  (External Ports (Total number : 4)
    (Name dout (Active Cycle 3 (Rate 50.00%)))
    (Name xi (Active Cycle 1 (Rate 16.67%)))
    (Name yi (Active Cycle 1 (Rate 16.67%)))
    (Name ou (Active Cycle 1 (Rate 16.67%)))
  )
  (FUs (Total Number : 1)
    (Name FU_1 (Active Cycle 2 (Rate 33.33%)))
  )
  (Muxes (Total number : 5)
    (Name x_1 (Active Cycle 3 (Rate 50.00%)))
    (Name x_2 (Active Cycle 2 (Rate 33.33%)))
    (Name x_3 (Active Cycle 2 (Rate 33.33%)))
    (Name x_4 (Active Cycle 2 (Rate 33.33%)))
    (Name x_5 (Active Cycle 2 (Rate 33.33%)))
  )
)
```

(b) statistics report

Figure 5.27 Evaluation and statistics reports

5.3.3 Incremental refinement design using behavioral synthesis

Behavioral synthesis can provide automatic solutions to circuit design. However better solutions are usually reached when designer's experience is exploited. Experienced designers often have a good idea of the architecture they aim at.

Therefore it is much wiser to exploit this knowledge and to allow it to be introduced as hints for the synthesis.

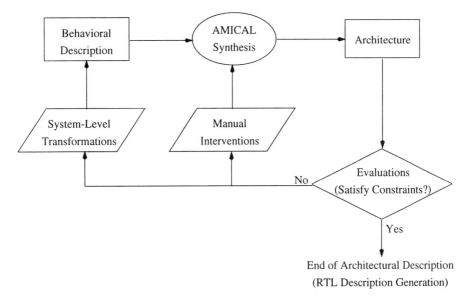

Figure 5.28 Incremental refinement based behavioral synthesis

Experiences using behavioral synthesis have shown that the specification and design of complex circuits generally follow the iterative scheme shown in figure 5.28. The designer starts with an idea of an architecture in mind. This will influence the way he will write the VHDL description. Starting from this initial specification, behavioral synthesis will produce an initial architecture; this may be automatically or manually. During this process the designer tries to understand the produced architecture and find how to improve it.

Architectural exploration can be executed at two different levels: a) by acting on the behavioral specification b) through manual interventions during scheduling and allocation steps and by exploiting the interactivity offered.

As stated within the architecture exploration steps, system level transformations and interactive modifications can be applied. System level modifications include system analysis and partitioning. These correspond to modifications of the behavioral VHDL code and of the functional unit library used for the AMICAL processing.

Hints for behavioral synthesis can also be introduced during the synthesis itself through interactive synthesis.

5.4 BEHAVIORAL SYNTHESIS IN THE DESIGN LOOP

This section discusses the main issues related to introducing behavioral synthesis in the design process. The main critical points are: debugging behavioral descriptions, linking with RTL and logic synthesis, mixing synthesizable and nonsynthesizable description, and validating the results of behavioral synthesis.

5.4.1 The design loop

Figure 5.29 shows a high-level synthesis based design flow with the corresponding verification process and design loops.

From the initial specification level, the circuit is manually partitioned and described at the behavioral level. The behavioral level description is simulated (validation B) to check the global functionality. Loop "B1" allows to debug the behavioral description.

From the behavioral level, a register transfer level architecture of the circuit is automatically generated using behavioral synthesis. The RTL description is simulated (validation R) to check the behavior at the clock cycle level. Loop "R1" is used to debug the behavioral description with regard to the synthesizable subset and the writing style accepted by behavioral synthesis. Loop "R2" is used to debug the directives and other inputs to behavioral synthesis for functionality such as the availability of the right components in the library.

From the register transfer level, the architecture of the circuit is mapped onto a gate level netlist and optimized with an RTL and logic synthesizer, according to manual directives. The gate level description is simulated (validation G) to check the delays. Loop "G1" debugs the behavioral description for lacking signal and variable initializations not detected during validation R. Loop "G2" debugs the directives for behavioral synthesis for performance optimization such as pipelining the control part and the data path. Loop "G3" debugs the directives for RTL and logic synthesis for performance optimization such as re-timing or pipelining inside the data path of the functional units. From the gate level, the circuit is then placed and routed.

Anatomy of a Behavioral Synthesis System Based On VHDL

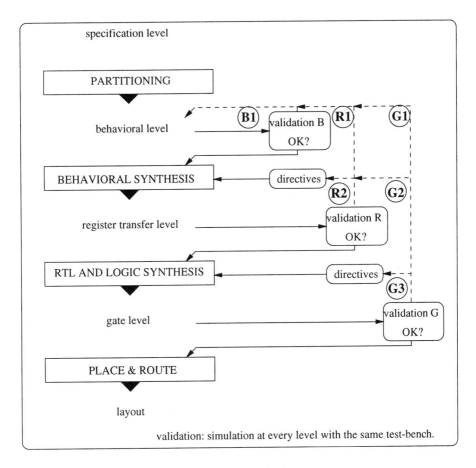

Figure 5.29 Design loops.

Because of the abstraction level and the amount of information handled, the behavioral simulation (validation B) is less time consuming than the RTL simulation (validation R) which is much faster than gate simulation (validation G). For the same reasons, behavioral synthesis is faster than RTL synthesis which is much faster than gate level synthesis.

A behavioral loop includes behavioral simulation, an analysis of the simulation results and a possible change of the behavioral description. The different tasks corresponding to different loops are given in table 5.2.

	Behav. sim.	Updating behavioral specifications	Behav. synt.	Updating behavioral directives	RTL sim.	RTL synt.	Updating RTL synt. directives	Gate level sim.
B1	X	X						
R1		X	X		X			
R2			X	X	X			
G1		X	X			X		X
G2			X	X		X		X
G3			X			X	X	X

Table 5.2 Design tasks in design loops

When looking to table 5.2 it becomes obvious that the behavioral loop (B1) is faster than RTL loops (R1,R2) which are much faster than Gate loops (G1,G2,G3).

On the other side it is easy to show that each behavioral loop may avoid several RTL loops and an RTL loop may avoid several gate loops. In fact bugs are easier to detect and to debug at the higher levels.

It is clear then that the use of behavioral synthesis induces a large reduction in the design time. Even when behavioral synthesis is not used several designer prefer to start with a behavioral description in order to fix the specification of the design. This allows to have several behavioral loops before starting RTL design.

However in practice, when behavioral design is not used, the behavioral description is not updated during the design process. In this case, the RTL description is generated manually and when bugs are detected they are not reported in the behavioral description. Although the design starts with a behavioral description, after few iterations in the design process the behavioral description becomes obsolete and the RTL description becomes the reference model.

When behavioral synthesis is used, the lower level models are generated automatically. In this case the behavioral description remains the reference model during all the design process.

5.4.2 Links with RTL and logic synthesis

Behavioral synthesis aims at bridging the gap between high level specification and register transfer or logic level tools. The output descriptions of behavioral synthesis need to be compatible with existing tools. Register transfer level synthesis tools usually accept only a subset of VHDL descriptions at this level. Moreover special writing styles need to be satisfied by the specifications for good synthesis results. A structural writing style gives descriptions fully synthesizable at the register transfer level. The architecture generated is given as a set of interconnected modules, mainly: the controller, which is given as a finite state machine defined at the clock cycle and the datapath, which is described using a set of simple interconnected components.

So, AMICAL high-level synthesis provides a VHDL description that can feed any RTL and logic synthesizer admitting the standard language VHDL. The RTL and logic synthesis is modular and hierarchical. It begins with the synthesis of the allocated functional units belonging to the library of the architectural synthesizer. Then the data path is synthesized.

The controller is submitted to logic synthesis. This step performs the optimization of the finite state machine automatically generated by the scheduler of the HLS tool.

5.4.3 Mixing synthesizable and nonsynthesizable description

At the behavioral level the VHDL architecture may be composed of: a behavioral part (one or more processes), a structural part and a data flow part. Only part of this description will be considered by the behavioral synthesis. The other parts are not synthesizable. Figure 5.30(a) shows a typical design flow using behavioral synthesis where the behavioral description includes nonsynthesizable parts. The synthesizable part is used by behavioral synthesis in order to produce an architecture made of a datapath and a controller. This architecture has to be connected to the nonsynthesizable part when generating the RTL model.

The listed nonsynthesizable parts and the synchronization block are added as glue cells or connections through a specific synthesis step called architecture personalization. This step makes use of a script called personalization file in order to connect the synthesizable part and the rest of the system.

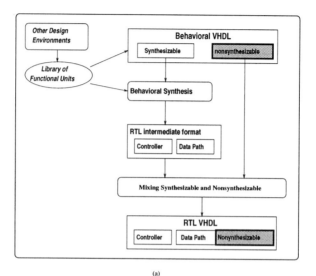

Figure 5.30 Mixing synthesizable and nonsynthesizable parts

5.4.4 Validation of the results of behavioral synthesis

The validation of the specification at the different synthesis steps is a critical stage when behavioral synthesis is used. As stated earlier (figure 5.29) the design loop may include several validation steps acting at different abstraction levels. This problem gets harder when the behavioral description is mixed with nonsynthesizable parts and even much harder when the design methodology imposes the use of a unique testbench for all the validation level. The use of different testbenches for different abstraction levels requires a nonacceptable extra effort and leads to incoherent situations if the different testbenches are not equivalent. The use of a single testbench allows to check that the global behavior of the circuit is the same whatever the level of abstraction. In this case the same test case will be used for the validation of the behavioral specification, the RTL model produced by behavioral synthesis and the gate level description produced by RTL synthesis. But this approach imposes lots of restrictions as it will be explained.

Figure 5.31 shows the different VHDL models that may be used for the validation during the design loop. We assume that we will use a unique testbench. There will be at least three models of the part synthesized using behavioral synthesis. These are called S1, S2 and S3. When nonsynthesizable parts are used, they may produce several intermediate models that have to be checked, these are called N1...Nn (n being the number of intermediate levels).

During the design process several cases of simulation will be needed. Each simulation will act on the testbench, a model of the synthesized part and a VHDL model of the non synthesized part. This may produce 3 x n cases of simulation. As explained in section 5.4.3, none of these is useless. Each may point out some problems and detect errors in the specification or in the synthesis process.

However, the application of the above mentioned methodology for simulation using a single testbench and mixing synthesizable ports and nonsynthesizable parts imposes lots of restrictions on the VHDL writing style and on the synthesis process. This comes from the fact that different abstraction levels use different time concepts (see 1.2) which may influence the timing of the I/O of different blocks. Additionally behavioral synthesis may even change the order of the I/O (see 3.2).

In order to be able to run all the above mentioned simulation cases, the communication between the three parts of the model (testbench, synthesizable part and nonsynthesizable parts) should be nonsensitive to the transformation allowed during behavioral synthesis. This may induce the use of complex communication procedures like handshaking and shared memory.

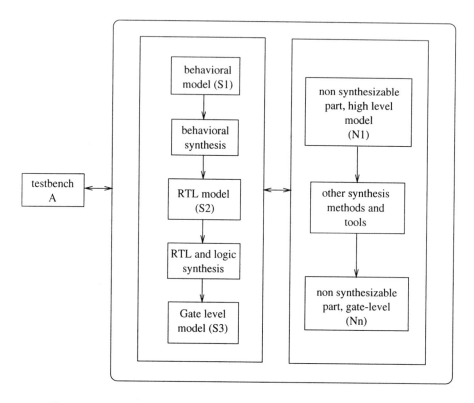

Figure 5.31 Different models used for the validation of the design and the design process

5.5 SUMMARY

This chapter introduces AMICAL, a VHDL behavioral synthesis tool allowing design reuse. The main design steps and design models were detailed.

The synthesis process is composed of six steps: compilation of the VHDL description, global scheduling and building of the behavioral FSMC, functional unit allocation and binding, micro-scheduling, connection allocation and controller/datapath generation.

AMICAL makes use of a behavioral synthesis model allowing to handle very large designs based on hierarchical specification and multi-level scheduling. The basic idea behind this model is that a complex system is generally composed of a set of subsystems performing specific tasks. A high-level specification of such

Anatomy of a Behavioral Synthesis System Based On VHDL   199

a system needs only to describe the sequencing of these tasks, consequently the coordination of the different subsystems. Each subsystem corresponding to a functional module designed (or selected) to perform a set of specific modules, is modeled as a unit executing a set of complex operations. Therefore the behavioral specification may be seen as a coordination of the activities of the different subsystems. The decomposition of a system specification into a global control and detailed tasks allows to handle very complex design through modularity.

This model is supported by the intermediate format (FSMC) and the target architecture. The application of AMICAL for hierarchical and modular application of the behavioral synthesis will be discussed in the next two chapters.

6

CASE STUDY: HIERARCHICAL DESIGN USING BEHAVIORAL SYNTHESIS

This chapter illustrates hierarchical design and design reuse at the behavioral level using a design example : a PID (Proportional Integral Derivative). Part of this design was already discussed in chapter 4. For a better understanding of the chapter we left them here also.

After an introduction of the PID application we recall the partitioning step discussed in chapter 4. A hierarchical solution is then implemented using behavioral synthesis. Behavioral synthesis produces a subsystem (a fixed-point unit) which is used in a second step as a functional unit by AMICAL for the synthesis of the full system.

6.1 THE PID

A PID controller usually applies a control function to an analog input and generates an analog output. This kind of device is generally implemented as an analog one. The use of a digital solution allows to have an integrated design. The input signal measures a process condition, while the output signal causes an actuator to either maintain or change the process conditions.

The PID used in this paper forms part of a speed control system detailed in [DPA91]. The speed control system includes an ALU which performs elementary and logic operations, and memories to store the state variables and coefficients. The speed reference is supplied by the host computer and the PID speed algorithm is executed each time a rotor position change has occurred.

The control loop making call to the PID is shown in figure 4.9. The controller device performs the 3 functions: speed control, current control and communication with a host computer.

The PID is executed by the processor once every n position interrupts. It is assumed that before execution, the motor parameters and control coefficients have already been loaded in the main processor. The algorithm calculates a current reference (*Iref*) as a linear expression of the rotational speed error (*Ek*), its time integral $\int (Ek) dt$ and its rate of change (*dEk/dt*).

6.2 SPECIFICATIONS

The PID algorithm is given by:

$$Irefk <= (Kp * Ek) + Ki * \int (Ek)dt + Kd * dEk/dt$$

where Kp, Ki, Kd are constants and Ek is the error change. However only the close approximation given as:

$$Irefk <= (Kp * Ek) + Ki * \sum (Ek * \delta T) + Kd * \Delta(Ek)/\delta T$$

will be developped in order to be synthesized at the behavioral level by AMICAL, for digital implementation. The integral is approximated by a sum of products while the derivative is replaced by successive position interrupts. The description of the main loop is given by figure 6.1.

```
1   while (HostInterrupt = '0') loop
2     wait until (PositionChange = '1');
3     -- N is the time between successive invocations of the algorithm
4     Fk := 1/N;           -- Calculation of speed
5     Ek_1 := Ek;          -- Storing speed error for next time
6     if (Fsignin = '0')
7     then Ek := Fref - Fk;
8     else Ek := Fref + Fk;
9     end if;
10    Dek := Ek - Ek_1;  -- Speed error change <- Ek - Ek_1
11    Irefk := Kp * Ek;  -- Irefk <- Kp*Ek
12    Temp := Dek * Fk;  -- Temp <- Dek*Fk, or Dek*(1/DeltaT)
13    Temp := Temp * Kd; -- Temp <- Kd*Dek*(1/DeltaT), or Kd*dEk/dt
14    Irefk := Irefk + Temp; -- Irefk <- Kp*Ek + Kd*dEk/dt
15    Temp := Ek * N;    -- Temp <- Ek*DeltaT
16    Ik := Ik + Temp;   -- Ik <- Ik + Ek*DeltaT, or Sum(SpeedError*DeltaT)
17    Temp := Ik * Ki;   -- Temp <- Ki * Sum(SpeedError*DeltaT)
18    Irefk := Irefk + Temp; -- Kp*Ek + Kd*dEk/dt + Ki*Integral(Ek*dt)
```

```
19    Irefkout <= Irefk;
20  end loop;
end behavior;
```

Figure 6.1 PID algorithm (VHDL extract)

6.3 SYSTEM-LEVEL ANALYSIS AND PARTITIONING

The goal of the system-level analysis and partitioning step is to structure the description in order to allow hierarchical specification and component reuse. The result of such step is a behavioral description and its corresponding functional unit library. In fact, the above PID model makes use of complex operations (*, /) that are not in the standard library. We will see hereafter how the components to be used in the design will be chosen and the impact of such choice on the behavioral description produced. This step is performed manually.

The only data types required by the speed control algorithm are numeric. Reals are used to represent the values of state variables such as current, frequency, etc., while integers are used for time intervals and clock frequency constants. The decision was made to use fixed-point arithmetic for numerical calculations on reals, thus avoiding the design of floating point hardware. Two's complement representation will be used for both integer and fixed-point real numbers. The word length is 32 bits. Fixed-point reals have a 12 bit integer part and 20 bits for the binary point part.

6.4 HIERARCHICAL DESIGN

The partitioning of the PID was discussed earlier in section 4.3. For the rest of this chapter, the design of the hierarchical solution figure 6.2 (or figure 4.10(c)) will be detailed. For the synthesis of the speed control design example, we will proceed in a hierarchical way. The fixed-point unit will be synthesized first and the results will be used as a functional unit for the design of the whole PID processor.

The full design process is shown in figure 6.3. The AMICAL process is run twice; once for the fixed-point unit synthesis and a second time for the PID synthesis. Starting from a behavioral description and a library of functional units, AMICAL produces a register transfer level description. The library of

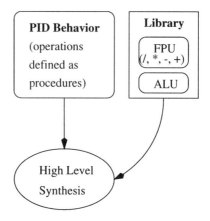

Figure 6.2 Hierarchical PID design

functional units used for the fixed-point unit synthesis is made up of an ALU which executes either addition or subtraction, and a shifter (undertaking right-shift as well as left-shift operations).

6.5 DESIGN FOR REUSE OF THE FIXED-POINT UNIT AS A BEHAVIORAL COMPONENT

As stated above, the fixed-point unit (fpu) executes addition, subtraction, multiplication as well as division. The multiplication algorithm selected is an add-and-shift based one, while the divisor (reciprocal as the numerator is always equal to one) makes use of the restoring division algorithm [MPC90].

The goal is to design the fixed-point unit as a functional unit. As explained in chapter 4, several views of the component should be defined. All these views are inter-related. An important step when defining new functional units is the definition of the behavioral interface and of the operations executed by the functional unit.

The execution time for division and multiplication are data-dependent and thus not fixed. The fixed-point unit should then provide a protocol in order to allow the use of these operations through actions with fixed execution time. However addition and subtraction need an exact number of clock cycles to execute.

Case study: hierarchical design using behavioral synthesis

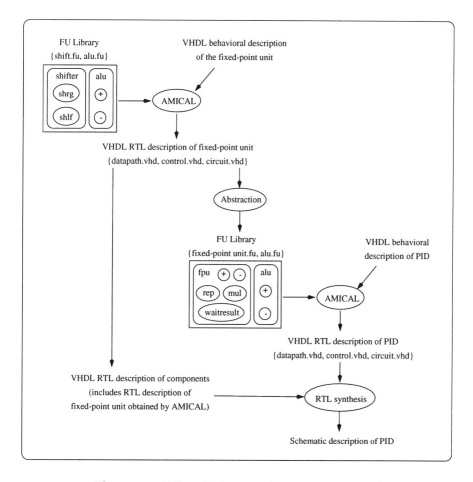

Figure 6.3 Full synthesis process (from behavior to gate)

When using these algorithms, the multiplication takes 126 or 127 (according to the parity) clock cycles for execution. For the same fixed-point unit, the execution time for the reciprocal can vary between 82 and 102 according to the input values, assuming that the fixed-point unit is synthesized with limited resources (1 shifter and 1 ALU (+ and -)). Therefore one of the main problems encountered during the fixed-point unit design abstraction was to describe the execution scheme of the complex operations: multiplication and division (reciprocal).

We thus decided to use a two-step protocol for the execution of both division and multiplication. Each of these 2 complex functions has been decomposed into 2 procedures. The first step is the operation call with the corresponding parameters. This step starts the computation (multiplication or division). During the computation the fixed-point unit will be blind to all external commands. This state is signaled through a low value on the signal *"done"*. The second step involves the *"waitresult"* operation which enables the result to be recovered. Both operation call and *"waitresult"* are described as one-clock-cycle operations in the high-level description of the operator in the library of functional units. The behavior of the fixed-point unit can be summarized by:

```
entity fixedpointunit is
port (A, B: in integer;        -- input values
      com: in integer;         -- operation asked
      sel: in bit;             --enable signal
      Z: out integer;          -- output value
      done: out bit );         -- output validation signal
end fixedpointunit;
architecture behavior of fixedpointunit is
begin
  process
    variable tmp: integer;         -- internal global variable; buffer
    variable input1, output2: integer;-- input value buffers
    procedure mul(A,B: in integer) is
    begin
      -- shift and add algorithm; tmp:= A * B;
    end mul;
    procedure rep(A: in integer)is
    begin
      -- restoring division algorithm; tmp:= 1/A;
    end rep;
  begin
    wait until sel='1';
    case com is
        when 1 => -- reciprocal_call
                  done <= '0'; input1 := A; rep(input1);
        when 2 => -- multiplication_call
                  done <= '0'; input1 := A;
                  input2 := B; mul(input1,input2);
        when 3 => -- "+"
                  Z = A + B;
        when 4 => -- "-"
                  Z <= A - B;
        when 5 => -- waitresult
                  Z <= tmp; done <= '1';
    end case;
  end process;
end behavior;
```

Figure 6.4 Fixed-point unit description

Case study: hierarchical design using behavioral synthesis   207

6.6 ABSTRACTION FOR REUSE

In order to use the fixed-point unit as a behavioral component, we need to perform an abstraction step that consists in extracting informations necessary for high-level synthesis. Three different views of the fixed-point unit interface are shown in figure 6.5. Some pieces of information can be obtained only after the high-level design of the fixed-point unit. This includes the number of cycles needed for the execution of each operation.

The behavioral view includes 5 operations, making use of 2 input and 2 output parameters ((A, B) and (Z, done) respectively). The protocol for parameter exchange is detailed in the synthesis view.

IMPLEMENTATION VIEW

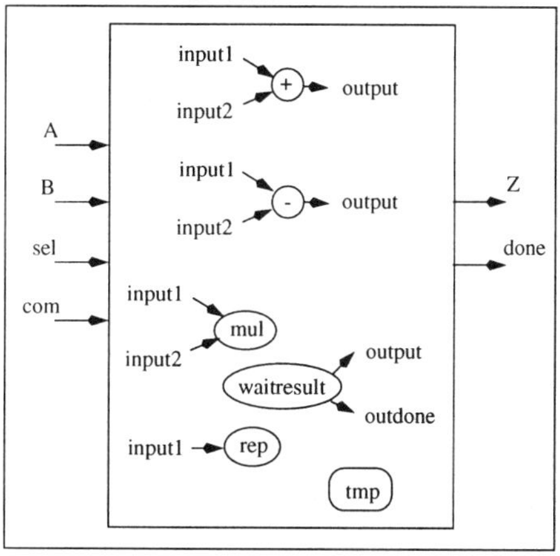

SYNTHESIS VIEW
```
(FU fpu
 (AREA 30000)(WIDTH 120)(HEIGHT 100)
 (PARAMETER (DataIn A B)(DataOut Z done)
 (CONNECTOR
  (DataIn input1 (BIT 32) input2 (BIT 32))
  (DataOut output (BIT 32) outdone (BIT 1))
  (ControlIn sel (BIT 1) com (BIT 3))
 (OpType + (commutative A B)
  (Cycle 1 (Transfer A input1)
   (Transfer B input2) (Transfer output Z)
   (active sel 1) (active com 3)))
 (OpType -
  (Cycle 1 (Transfer A input1)
   (Transfer B input2) (Transfer output Z)
   (active sel 1) (active com 4)))
 (OpType mul (commutative A B)
  (Cycle 1
   (Transfer A input1) (Transfer B input2)
   (active sel 1) (active com 2)))
 (OpType rep
  (Cycle 1 (Transfer A input1)
   (active sel 1) (active com 1)))
 (OpType waitresult
  (Cycle 1
   (Transfer output Z) (Transfer outdone done)
   (active sel 1) (active com 5)))))
```

BEHAVIORAL VIEW

procedure "+" (A,B:in integer; Z:out integer);
procedure "-" (A,B:in integer; Z:out integer);
procedure mul (A,B: in integer);
procedure rep (A: in integer);
procedure waitresult (Z:out integer; done:out bit);

Figure 6.5   Fixed-point unit interface

6.7 DESIGN REUSE

The 2-step protocol applied for the multiplication operation can be summarized by figure 6.6. The multiplication call is made using as input values Kp and Ek. When the multiplication is over, the result is placed in a register and later the procedure waitresult brings back this multiplication result. In order to point out when the multiplication result is ready, a validity signal called "*done*" is used. This results in writing a multiplication call followed by a loop of *waitresult* until the result is ready. The result of the multiplication is placed in *Irefk* according to the initial specification and the validity signal is affected to done. For component reuse, the multiplication has thus been described as a 2-function operation. Such a decomposition can be applied to any complex operations. However during high-level synthesis and more precisely during allocation, all the sub-operations need to be undertaken by the same functional unit as they share data. In the case of AMICAL, this is ensured automatically in most cases. When more than one functional unit able to execute a procedure using local data is allocated, manual operations may be needed to adjust the results of automatic synthesis. However manual operations may be avoided by constraining the writing style of the VHDL description.

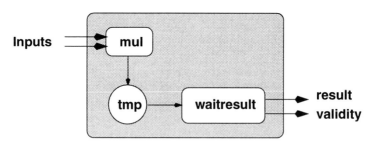

Figure 6.6 2-step protocol for multiplication

The use of the fixed-point unit as a component of the functional unit library results in expanding the multiplication or division call according to the protocol of the fixed-point unit. The full VHDL description after partitioning is given in figure 6.7.

```
package op_pkg is
   subtype int3bit is integer range 0 to 7;
end op_pkg;

package synchro is
   function rising_edge(signal sig:bit) return boolean;
end synchro;
```

Case study: hierarchical design using behavioral synthesis   209

```
package body synchro is
   function rising_edge(signal sig:bit) return boolean is
   begin
     return (sig'event and sig='1' and horloge'last_value='0');
   end rising_edge;
end synchro;

use work.synchro.all;
use work.op_pkg.all;
------------------------------------------------------------------------
entity pid is
port
  (clock         : in bit;
   reset         : in bit;
   Fsignin       : in bit;         -- Direction of travel
   HostInterrupt : in bit;         -- On signal
   PositionChange: in bit;         -- New position validation
   Irefkout      : out integer);   -- Output : motor drive current
end pid;
------------------------------------------------------------------------
architecture behavior of pid is

  component fu_fpu
  port
    (clock   : in bit;
     reset   : in bit;
     input1  : in integer;
     input2  : in integer;
     sel     : in bit;
     com     : in int3bit;
     output  : out integer;
     outdone : out bit);
  end component;
  signal sig_in1, sig_in2, sig_out : integer;
  signal sig_sel, sig_done : bit;
  signal sig_com : int3bit;

begin

  Inst_FU : fu_fpu
  port map(clock   => clock,
           reset   => reset,
           input1  => sig_in1,
           input2  => sig_in2,
           sel     => sig_sel,
           com     => sig_com,
           output  => sig_out,
           outdone => sig_done);

  process
    variable Ik, Ek, Kp, Ki, Kd, Fref: integer;
    variable N, Fk, Ek_1, Dek, Irefk, Temp: integer;
    variable done: bit;
    procedure mul(a,b:in integer) is
    begin
```

```
    sig_in1 <= a; sig_in2 <= b;
    sig_sel <= '1'; sig_com <= 2;
    wait until rising_edge(clock);
  end mul;
  procedure rep(a:in integer) is
  begin
    sig_in1 <= a;
    sig_sel <= '1'; sig_com <= 1;
    wait until rising_edge(clock);
  end rep;
  procedure waitresult(x:out integer; y:out bit) is
  begin
    x := sig_out; y := sig_done;
    wait until rising_edge(clock);
  end waitresult;

  type ROM is array(0 to 4) of integer;
  variable val_rom : ROM := (2 * 2**20, 6 * 2**20, 3 * 2**20,
                             4 * 2**20, 25 * 2**20);

  procedure getconstKp(x:out integer) is
  begin
    x := val_rom(0);
  end getconstKp;
  procedure getconstKi(x:out integer) is
  begin
    x := val_rom(1);
  end getconstKi;
  procedure getconstKd(x:out integer) is
  begin
    x := val_rom(2);
  end getconstKd;
  procedure getFref(x:out integer) is
  begin
    x := val_rom(3);
  end getFref;
  procedure getN(x:out integer) is
  begin
    x := val_rom(4);
  end getN;
-------------------------------------------------------------------
begin
  getconstKp(Kp);      -- Constant
  getconstKi(Ki);      -- Constant
  getconstKd(kd);      -- Constant
  getFref(Fref);       -- Required speed
  Ik := 0; Ek := 0;
  wait until (HostInterrupt = '0');
  while (HostInterrupt = '0') loop
    wait until (PositionChange = '1');
    getN(N);   -- Time between successive invocations of algo
    rep(N); --Fk := 1/N;-- Calculate speed
    Ek_1  := Ek;         -- Store Speed error for next time
    --wait until (done = '1');
    waitresult(Fk,done);
    while (done /= '1') loop waitresult(Fk,done); end loop;
```

Case study: hierarchical design using behavioral synthesis   211

```
      if (Fsignin = '0')
      then Ek := Fref - Fk; else Ek := Fref + Fk; end if;
      mul(Kp, Ek);         -- Irefk := Kp*Ek
      Dek := Ek - Ek_1;    -- Speed error Change := Ek - Ek-1
      --wait until (done = '1');
      waitresult(Irefk,done);
      while (done /= '1') loop waitresult(Irefk,done); end loop;
      mul(Dek, Fk);        -- Temp := Dek*Fk, or, Dek*(1/DeltaT)
      --wait until (done = '1');
      waitresult(Temp,done);
      while (done /= '1') loop waitresult(Temp,done); end loop;
      mul(Temp, Kd);       -- Temp := Kd*Dek*(1/DeltaT), or, Kd*dEk/dt
      --wait until (done = '1');
      waitresult(Temp,done);
      while (done /= '1') loop waitresult(Temp,done); end loop;
      Irefk := Irefk + Temp;-- Irefk := Kp*Ek + Kd*dEk/dt
      mul(Ek, N);          -- Temp := Ek*N, or, Ek*DeltaT
      --wait until (done = '1');
      waitresult(Temp,done);
      while (done /= '1') loop waitresult(Temp,done); end loop;
      Ik := Ik + Temp;     -- Ik := Sum(SpeedError*DeltaT)
      mul(Ik, Ki);         -- Temp := Ki*Sum(SpeedError*DeltaT)
      --wait until (done = '1');
      waitresult(Temp,done);
      while (done /= '1') loop waitresult(Temp,done); end loop;
      Irefkout <= Irefk + Temp;-- Kp*Ek + Kd*dEk/dt + Ki*Int(Ek*dt)
    end loop;
  end process;

end behavior;
```

Figure 6.7 PID algorithm described in VHDL

6.8 THE BEHAVIORAL SYNTHESIS PROCESS

As explained in figure 6.3, the architectural synthesis of both fixed-point unit and PID are realized by AMICAL. The results obtained are shown in figure 6.8a and figure 6.8b respectively.

The result of the synthesis of the fixed-point unit will be used twice:

1. It will be used to create the corresponding behavioral component that will be used for the synthesis of the PID.

2. It will be used during the logic synthesis of the PID as the required structural description of the functional unit: fixed-point unit.

The synthesis of the fixed-point unit has thus been made first, it produced an architecture composed of a datapath and a controller. Figure 6.8a shows the datapath generated by AMICAL.

The functional unit library used as input for the PID synthesis includes the fixed-point unit developed above and an ALU that may execute addition and subtraction. The behavioral description of the PID is written according to the protocol used by the fixed-point unit.

The synthesis of the PID produces an architecture where the controller is a 22-state and 33-transition finite state machine. The datapath obtained after some interactive architectural transformations is made up of 3 functional units. Within the PID datapath, one of the components is an instance of the fixed-point unit compiled previously. This is shown in figure 6.8b; an instance of the fixed-point unit is represented by FU_2.

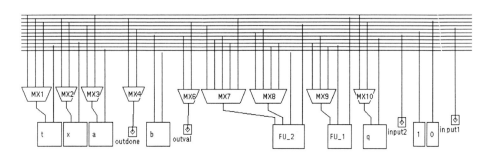

Figure 6.8a Synthesis results of the fixed-point unit datapath with AMICAL

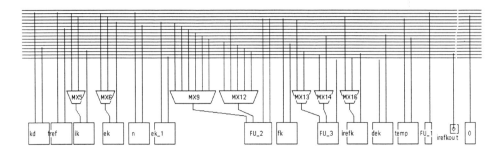

Figure 6.8b Synthesis results of PID datapath with AMICAL

The synthesis results obtained for the PID is composed of around 50000 transistors. The fixed- point unit itself is composed of more than 20000 transistors. The full design stands on 10.5 mm square when mapped onto a 0.8 CMOS technology using commercial logic synthesis and place and route tools.

6.9 SUMMARY

This chapter presented a design case where behavioral synthesis is applied through a hierarchical design. A fixed-point unit designed using behavioral synthesis is abstracted and reused as a functional unit for the behavioral synthesis of a PID system.

The scheme detailed above is powerful as it allows hierarchical design. This structured method enables the use of complex subsystems as functional units in the library. The major impact of the use of hierarchical synthesis on the design process is the decrease obtained in the synthesis time.

The higher the abstraction level and the greater the number of hierarchies, the smaller is the system specification. The computation time for the synthesis at the behavioral level is smaller and the behavioral description is simpler for modification and validation. Therefore on having the maximum number of iterations made at higher abstraction levels, the designers can accelerate the design process.

7

CASE STUDY: MODULAR DESIGN USING BEHAVIORAL SYNTHESIS

This chapter discusses the use of behavioral synthesis for the design of a complex system from its VHDL description.

This is illustrated by a window searching algorithm based system. The latter is a common application searching a specific *sequence* of words within a *string* of words. Such a system is usually designed using an architecture with a top controller managing three parallel subsystems: two memory modules and a coprocessor performing the search. Each subsystem is designed with high-level synthesis and includes a local controller. This chapter discusses the design of the subsystems and of the top controller described in VHDL using behavioral synthesis. The behavioral synthesis tool AMICAL is used throughout this chapter for illustration, but the behavioral description style may be adapted for other behavioral synthesis tools with VHDL inputs.

7.1 INTRODUCTION

A *Window Searching algorithm based System* (WSS) is a very common application. Its principle is similar to that of several algorithms searching a specific *sequence* of words in a *string* of words. Here follow three examples using this type of algorithm. The first one is the search of a gene (*sequence*) in a DNA string (*string*) [DiL88]. The second one is the translation of a sentence or a set of words (*sequence*) according to a dictionary of sentences (*string*). The third one is the motion estimation for the coding of a sequence of pictures, i.e. finding the position of a portion of the current picture (*sequence*) in a larger portion of the previous picture (*string*) [Wis89].

A typical model implementing a *WSS* algorithm is a specific application architecture made of a top controller and a datapath including two memory modules (respectively for *sequence* and *string*) and a coprocessor searching the most similar sequence in the *string*. Figure 7.1 shows a global view of such an architecture. This decomposition is made in order to allow the concurrent functioning of the three units. The memory modules may be filled while the coprocessor is executing the search algorithm. The top controller is in charge of coordinating the three components. The use of dual-port memories allows simultaneous read and write operations and therefore to fill them in parallel with the search process. Four issues have to be solved during the behavioral synthesis of such a design:

1. When analyzed, the system can be decomposed easily into two separate main subsystems: a Top Controller and coprocessors. How should the behavioral synthesis of such a global system be performed?

2. As described above, the datapath uses functional units (memory modules and a coprocessor) working concurrently. How does the behavioral synthesis allow such a parallelism?

3. To validate the behavioral description, accurate models (including set-up time, ...) of the functional units must be used. How can existing components such as RAM or coprocessors be reused at the behavioral level?

4. In order to deal with speed requirements, direct interconnections between functional units are needed (instead of being managed by the top controller). How are these local interconnections translated after the behavioral synthesis?

7.2 SYSTEM SPECIFICATION

This section describes the functionality, the memory and speed requirements, and the algorithm of the overall system *WSS*.

7.2.1 Functionality

The functionality of the Window Searching algorithm based System is to select, in an n-character *string*, a sequence of characters which is the closest to a

Case Study: Modular Design Using Behavioral Synthesis 217

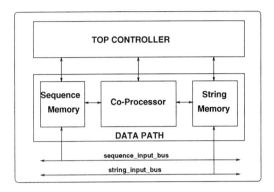

Figure 7.1 A global architecture of a Window Searching algorithm based System

specific *sequence* characters. We will assume that the *string* is made of 23 characters W_i ($0 \leq i \leq 22$). We also assume that the *sequence* is 8 character long Y_j ($0 \leq j \leq 7$). The *string* is decomposed into a set of bursts. In this case we assume that a burst is made of 8 characters. The *string* is then made of 3 bursts ($n = 3$). In many designs, the burst corresponds to the amount of data that can be loaded from the external memory. The *string* will be called the search window and the *sequence* will be called the current window. Both windows are loaded by bursts using two external buses (figure 7.1). Figure 7.2 shows a visual representation of the search process. The result consists of a motion vector varying from 0 to 15 and of the corresponding distortion between the specific *sequence* and the most similar sequence. The 24th character of the string is ignored as the motion vector size is limited. The motion vector X is defined by:

$$D_X = \min_{0 \leq x \leq 15} D_x$$

And the distortion is defined by:

$$D_x = \sum_{0 \leq i \leq 7} |Y_i - W_{i+x}|$$

7.2.2 Algorithm

Figure 7.3(a) shows a global view of the search algorithm: after an initialization step, the *sequence* is loaded. The *string* is loaded by n 8-character bursts. It

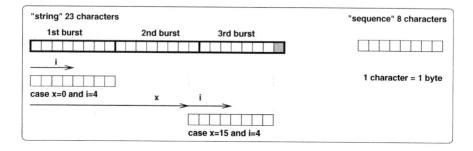

Figure 7.2 Visual representation of the search process

takes n-1 iterations. As long as the end of the *string* is not reached, a new burst is loaded. Once two consecutive bursts of the *string* are loaded, the search process may start. Each task in figure 7.3(a) is labeled with a duration (t_i). The time needed for the completion of an iteration of the whole algorithm is:

$$T_a = t_1 + t_2 + t_3 + (n-1).(t_3 + t_4)$$

where n is the number of bursts needed to fill a *string*.

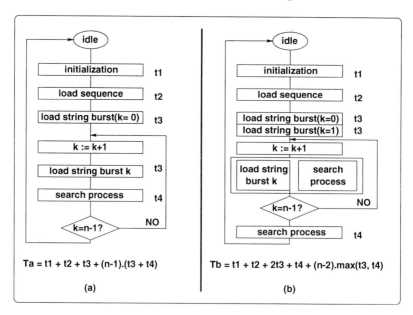

Figure 7.3 Search algorithms

7.2.3 Memories

This search algorithm is based on two read/write memory areas for *sequence* and *string*. As they are read simultaneously by the coprocessor, we will store them in two different RAMs. The *string* and the *sequence* are loaded from an external memory using an asynchronous protocol. As stated above they are filled burst by burst. In order to allow concurrence between the top controller and the memory filling, local controllers are needed. So, two modules consisting each of a RAM and a local controller are used: *mem_sequence* encapsulates an 8x8 bits RAM containing the *sequence* and *mem_string* encapsulates a 24x8 bits RAM containing the *string*.

7.2.4 Parallelism

Two types of parallelism are possible: between the loading of the two memories before the inner loop and between the search processing and the load of the next burst to be processed.

The parallelism between the filling of both memories depends on how the burst arrives. We assume that each burst is available after a request/acknowledge protocol on an external bus. As shown in figure 7.1, two external buses may be used: one for the *sequence* burst and one for the *string* bursts. In this case, both memories can be written in parallel. But in order to reduce the number of external connections, we will use a single bus (*Data_In*). So, the memories cannot be loaded in parallel (section 7.3.2).

The parallelism between the search processing and the load of the next burst to be processed supposes that *mem_string* contains a RAM, allowing reading and writing at the same time. This accounts for the choice of a dual port RAM.

The sequencing and the parallelism of the different tasks are shown in a new version of the algorithm in figure 7.3(b). The time (T_b) for the global execution is shorter for this solution.
$$T_b = t_1 + t_2 + 2t_3 + (n-2).max(t_3, t_4)$$

7.3 SYSTEM PARTITIONING

This section sets out the system partitioning and the abstract architecture of the circuit according to the system specification.

7.3.1 Partitioning

The partitioning of the *WSS* results naturally from the top system specification. In fact three functional units are needed for this system: the encapsulated memories, *mem_sequence* and *mem_string*, and a coprocessor that performs the search algorithm. These components may already be generated by a behavioral synthesizer or by another design environment. Due to their complexity the memory controllers and the coprocessor may be generated by behavioral synthesis. So, four behavioral descriptions are required, corresponding to *mem_sequence*, *mem_string*, the coprocessor and the overall system. The high-level synthesis of the global architecture and the abstraction models of the components are described in the next sections.

7.3.2 Abstract architecture

From the partitioning, an abstract architecture of the *WSS* can be deduced. The system is viewed as an assembly of subsystems co-ordinated by a top controller. Each task of figure 7.3(b) can be executed by one of the three subsystems: "load sequence" by *mem_sequence*, "load string" by *mem_string* and the two modes of "search process" by a coprocessor. For the coprocessor, before the last burst processing ($k < n$), the partial results must be stored until the final search on the last burst ($k = n$). Figure 7.4 shows the resulting abstract architecture of the *WSS*. The functional units are interconnected: indeed the memories *mem_sequence* and *mem_string* receive external inputs (*Data_In*) and the coprocessor loads these data from the memories.

7.3.3 Design flow for modular use of high-level synthesis

This section applies the methodology presented in section 4.3 for modular use of the behavioral synthesis tool AMICAL. The *WSS* system is composed of three blocks. Because of their complexity and in order to make easier the task of the top controller, we decided that each block contains its own local controller. The global system, as well as its blocks, will be described at the behavioral level and submitted to high-level synthesis. Figure 7.5 shows the corresponding design flow.

The components are designed in a first step. In this case we will use AMICAL for the design of the two memory modules and the coprocessor. In a second

Case Study: Modular Design Using Behavioral Synthesis 221

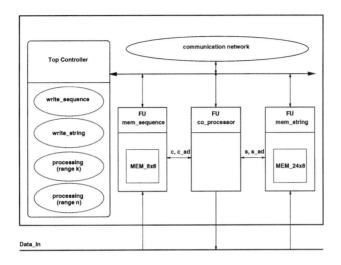

Figure 7.4 Abstract architecture of the *WSS*

step, the processor is abstracted in order to be reused for the design of the whole system. Finally the top controller is synthesized using AMICAL.

Sections 7.4 and 7.5 describe respectively the components and the global *WSS* designed with AMICAL.

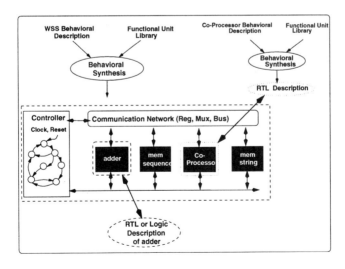

Figure 7.5 Modular use of high-level synthesis

7.3.4 VHDL description style

During this design we decide to use a full validation methodology based on a unique testbench and allowing to simulate all the parts of the system at different abstraction levels as introduced in section 5.4.4.

This choice imposes several restrictions on the VHDL writing style. The writing style for each component is imposed by its connection to the external world. For the components we mixed the cycle fixed I/O mode and the behavioral state fixed I/O mode. Since the memory blocks interact with the external data bus and the coprocessor, at the clock-cycle level, all the parts related to this communication are described using the cycle fixed I/O mode. The rest of the behavior, related to communication with the top controller, is described using the behavioral state fixed I/O mode. The top controller is also described using a behavioral state fixed I/O mode.

Like most behavioral synthesis tools, AMICAL interprets the wait statement as a synchronous transition even if the sensitivity list of this wait statement is not related to a clock. In order to have equivalent synchronous descriptions before and after synthesis, the writing style imposes synchronization at the behavioral level. This synchronization is made using the *rising_edge* IEEE function applied to the *clock*. One can note that the wait statement may include complex boolean expressions using other signals than the clock.

All the components and the top controller are connected to a global reset signal called *reset*.

The following packages are used for subtype and component declarations. Subtype declarations are needed for AMICAL synthesis in order to distinguish the different widths of the signals when using vector types.

```
1   LIBRARY ieee;
2     USE ieee.std_logic_1164.ALL;
3     USE ieee.std_logic_arith.ALL;
4
5   PACKAGE pkg_types IS
6
7           SUBTYPE bit1    IS std_ulogic;
8
9           SUBTYPE bit2    IS std_ulogic_vector ( 1 DOWNTO 0);
10          SUBTYPE bit3    IS std_ulogic_vector ( 2 DOWNTO 0);
11          SUBTYPE bit4    IS std_ulogic_vector ( 3 DOWNTO 0);
12          SUBTYPE bit5    IS std_ulogic_vector ( 4 DOWNTO 0);
13          SUBTYPE bit8    IS std_ulogic_vector ( 7 DOWNTO 0);
14          SUBTYPE bit11   IS std_ulogic_vector (10 DOWNTO 0);
```

```
15
16              SUBTYPE bit2_r  IS std_logic_vector ( 1 DOWNTO 0);
17              SUBTYPE bit3_r  IS std_logic_vector ( 2 DOWNTO 0);
18              SUBTYPE bit4_r  IS std_logic_vector ( 3 DOWNTO 0);
19              SUBTYPE bit5_r  IS std_logic_vector ( 4 DOWNTO 0);
20              SUBTYPE bit8_r  IS std_logic_vector ( 7 DOWNTO 0);
21              SUBTYPE bit11_r IS std_logic_vector (10 DOWNTO 0);
22
23   END pkg_types;
```

Figure 7.6 Package with subtype declarations

Figure 7.7 gives the component declaration package. The components *mem_sequence*, *mem_string* and coprocessor are respectively declared at lines 38-52, 85-100 and 102-118 of the package *pkg_components*. Each memory block contains a local controller managing a dual-port ram (l.26-36 for *mem_sequence* and l.73-83 for *mem_string*). In fact each dual port ram encapsulates itself the true ram component (l.7-24 for dpram_8x8 and l.54-71 for dpram_24x8) with some glue.

```
1    LIBRARY ieee;
2       USE ieee.std_logic_1164.ALL;
3       USE work.pkg_types.ALL;
4
5    PACKAGE pkg_components IS
6
7      COMPONENT dpram_8x8
8        PORT(
9              q2        : OUT   bit8;
10             q1        : OUT   bit8;
11             d2        : IN    bit8;
12             d1        : IN    bit8;
13             a2        : IN    bit3;
14             a1        : IN    bit3;
15             oen2      : IN    bit1;
16             oen1      : IN    bit1;
17             wen2      : IN    bit1;
18             wen1      : IN    bit1;
19             csn2      : IN    bit1;
20             csn1      : IN    bit1;
21             ck2       : IN    bit1;
22             ck1       : IN    bit1
23           );
24     END COMPONENT;
25
26     COMPONENT mem_8x8
27       PORT(
28             q2        : OUT   bit8;
29             d1        : IN    bit8;
30             a2        : IN    bit3;
31             a1        : IN    bit3_r;
32             sel_read: IN    bit1;
33             sel_write: IN    bit1;
34             ck        : IN    bit1
```

```
35              );
36      END COMPONENT;
37
38      COMPONENT mem_sequence
39        PORT(
40              clk     : IN    bit1;
41              reset   : IN    bit1;
42              sel_read: IN    bit1;
43              c_sel   : IN    bit1;
44              c_req   : OUT   bit1;
45              c_ack   : IN    bit1;
46              c_valid : IN    bit1;
47              data_in : IN    bit8;
48              c_ad    : IN    bit3;
49              c       : OUT   bit8;
50              c_done  : OUT   bit1
51              );
52      END COMPONENT;
53
54      COMPONENT dpram_24x8
55        PORT(
56              q2      : OUT   bit8;
57              q1      : OUT   bit8;
58              d2      : IN    bit8;
59              d1      : IN    bit8;
60              a2      : IN    bit5;
61              a1      : IN    bit5;
62              oen2    : IN    bit1;
63              oen1    : IN    bit1;
64              wen2    : IN    bit1;
65              wen1    : IN    bit1;
66              csn2    : IN    bit1;
67              csn1    : IN    bit1;
68              ck2     : IN    bit1;
69              ck1     : IN    bit1
70           );
71      END COMPONENT;
72
73      COMPONENT mem_24x8
74       PORT(
75              q2      : OUT   bit8;
76              d1      : IN    bit8;
77              a2      : IN    bit5;
78              a1      : IN    bit5_r;
79              sel_read: IN    bit1;
80              sel_write: IN   bit1;
81              ck      : IN    bit1
82              );
83      END COMPONENT;
84
85      COMPONENT mem_string
86        PORT(
87              clk     : IN    bit1;
88              reset   : IN    bit1;
89              sel_read: IN    bit1;
90              s_sel   : IN    bit1;
```

Case Study: Modular Design Using Behavioral Synthesis 225

```
 91           burst    : IN   bit2_r;
 92           s_req    : OUT  bit1;
 93           s_ack    : IN   bit1;
 94           s_valid  : IN   bit1;
 95           data_in  : IN   bit8;
 96           s_ad     : IN   bit5;
 97           s        : OUT  bit8;
 98           s_done   : OUT  bit1
 99           );
100    END COMPONENT;
101
102    COMPONENT co_processor
103      PORT (
104           clk      : IN   bit1;
105           reset    : IN   bit1;
106           c        : IN   bit8;
107           s        : IN   bit8;
108           p_sel    : IN   bit1;
109           mode     : IN   bit1;
110           sel_read : OUT  bit1;
111           c_ad     : OUT  bit3;
112           s_ad     : OUT  bit5;
113           dmin     : OUT  bit11;
114           vector   : OUT  bit4;
115           done0    : OUT  bit1;
116           done1    : OUT  bit1
117           );
118    END COMPONENT;
119  END pkg_components;
```

Figure 7.7   Component declaration package

7.4 BEHAVIORAL SPECIFICATIONS OF SUBSYSTEMS

The first step of the modular design is the high-level synthesis of the subsystems to be reused as components. The three components: *mem_sequence*, *mem_string* and the coprocessor are synthesized. The complete behavioral VHDL description of each module is given in this section.

7.4.1 The coprocessor

The coprocessor is in charge of executing the comparison process as introduced in section 7.2.1. The coprocessor is connected to the two memories, to the external bus and to the top controller as shown in figure 7.8.

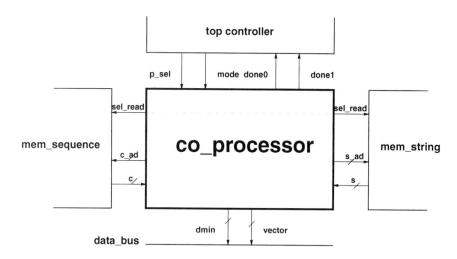

Figure 7.8 Connections of the *co_processor*

The communication protocol with the memories takes 2 cycles. During the first cycle, the coprocessor computes the addresses, selects the two memories in read mode (*sel_read*) and sends the addresses to the memories (*c_ad*, *s_ad*). During the second cycle the data from the memory is available (*s*, *c*).

The coprocessor is activated through the signals *p_sel* and *mode*. *p_sel* selects the coprocessor and *mode* fixes the running mode of the coprocessor.

The coprocessor has two running modes depending on the position of the *string* processed. The last iteration needs some specific computations. All the iterations except the last one are processed in the first mode. This mode is selected by *mode='0'* and *p_sel ='1'*. At the end of the computation, the coprocessor sets the signal *done0*. The last iteration is processed using a specific mode selected when *mode='1'* and *p_sel ='1'*. At the end of the computation, it sets *done1*. *done0*, *done1* and *p_sel* are external signals.

In addition to standard operations (+, incr, -), the coprocessor makes use of several specific computations mixing arithmetic operations, concatenations and type conversions. All these operations are gathered into a specific operator called *min_addr*. Each of these operations is executed as VHDL procedures (*assign*, *sel_min*, ...).

Case Study: Modular Design Using Behavioral Synthesis

The communication with the memories imposes that the input signals are stable on the rising edge of the clock. This timing constraint is hard to specify in timeless behavioral models. This comes from the fact that all assignments take zero delay and consequently during behavioral simulation all the signal assignments will happen on the rising edge of the clock. In order to avoid this problem all the input assignments of the memories are delayed using an after clause (*after 2 ns*). This style does not alter the behavioral style of the description since these procedures are not expanded during behavioral synthesis when using AMICAL.

Figure 7.9 shows the abstract architecture of the coprocessor. Each procedure, function or operation call requires a functional unit able to execute it and allows the allocation of the functional unit during behavioral synthesis with AMICAL. The procedures *assign*, *init_min* and *sel_min*, the function *dist* and the operator + are executed respectively by the functional units called *min_addr*, *alu* and *incr*.

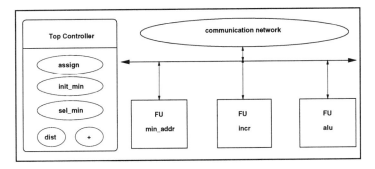

Figure 7.9 Abstract architecture of the *co_processor*

The VHDL behavioral description of the coprocessor is presented in figure 7.11. Its architecture is composed of a single process [1.25-108].

The process starts with *p_sel* for both modes [1.70], according to the protocol in figure 7.10. The initialization [1.73-84] depends on the mode. Then, there are two nested while loops: *loop_x* to scan all the vectors x [1.87-107] and *loop_i* for the distortion accumulation [1.89-101]. One iteration of the accumulation loop takes one clock cycle. For each RAM a new address must be assigned at each clock cycle to fetch one pixel, therefore there is a wait statement on the rising edge of the clock within the nested loops [1.93].

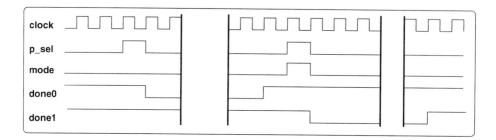

Figure 7.10 Activation protocol between the top controller and the coprocessor

```
1    LIBRARY ieee;
2      USE ieee.std_logic_1164.all;
3      USE ieee.std_logic_arith.all;
4      USE ieee.std_logic_signed.all;
5      USE work.pkg_types.all;
6
7    ENTITY co_processor IS
8    PORT (  clk      : IN   bit1;
9            reset    : IN   bit1;
10           c        : IN   bit8;
11           s        : IN   bit8;
12           p_sel    : IN   bit1;
13           mode     : IN   bit1;
14           sel_read: OUT  bit1;
15           c_ad     : OUT  bit3;
16           s_ad     : OUT  bit5;
17           dmin     : OUT  bit11;
18           vector   : OUT  bit4;
19           done0    : OUT  bit1;
20           done1    : OUT  bit1);
21   END co_processor;
22
23   ARCHITECTURE behavior OF co_processor IS
24   BEGIN
25     PROCESS
26     VARIABLE  i           : bit3_r;
27     VARIABLE  x           : bit4_r;
28     VARIABLE  d, min      : bit11_r;
29
30     FUNCTION dist (c, s: IN bit8; d: IN bit11_r) RETURN bit11_r IS
31     VARIABLE val8  : bit8_r;
32     VARIABLE val11 : bit11_r;
33     BEGIN
34       val8  := to_stdlogicvector(c) - to_stdlogicvector(s);
35       val8  := abs(val8);
36       val11 := val8 + d;
37       RETURN(val11);
38     END dist;
39
40     PROCEDURE assign (i: IN bit3_r; x: IN bit4_r;
41              SIGNAL c_ad: OUT bit3; SIGNAL s_ad: OUT bit5) IS
```

```
42    BEGIN
43      c_ad <= to_stdulogicvector(i) AFTER 2 ns;
44      s_ad <= to_stdulogicvector(("00" & i) +
                                    ('0' & x)) AFTER 2 ns;
45    END assign;
46
47    PROCEDURE init_min (SIGNAL dmin: OUT bit11) IS
48    BEGIN
49      min  := "11111111000";
50      dmin<= "00000000000";
51    END init_min;
52
53    PROCEDURE sel_min (d: IN bit11_r; x: IN bit4_r;
54                       SIGNAL dmin   : OUT bit11;
55                       SIGNAL vector: OUT bit4) IS
56    BEGIN
57      IF unsigned(d) <= unsigned(min)
58      THEN
59           min      := d;
60           vector   <= to_stdulogicvector(x);
61      END IF;
62      dmin <= to_stdulogicvector(min);
63    END sel_min;
64
65   BEGIN
66     sel_read <= '0';
67     done0 <= '1';
68     done1 <= '1';
69
70     WAIT UNTIL p_sel='1' AND rising_edge(clk) AND reset='1';
71
72     IF mode='0' OR reset='0' THEN
73            i        := "000";
74            x        := "0000";
75            d := "00000000000";
76            init_min(dmin);
77            vector   <= "0000";
78            done0    <= '0';
79     ELSE
80            IF mode='1' THEN
81                   done1 <= '0';
82            END IF;
83     END IF;
84
85     WAIT UNTIL p_sel='0' AND rising_edge(clk) AND reset='1';
86
87     loop_x: WHILE reset='1' LOOP
88            d := "00000000000";
89            loop_i: WHILE reset='1' LOOP
90                   sel_read <= '1';
91                   assign(i, x, c_ad, s_ad);
92
93                   WAIT UNTIL rising_edge(clk);
94
95                   sel_read <= '0';
96                   d := dist(c,s,d);
```

```
97                    i := i + 1;
98                    IF i=0 THEN
99                         EXIT loop_i;
100                   END IF;
101              END LOOP loop_i;
102              sel_min(d, x, dmin, vector);
103              x:= x + 1;
104              IF (x=0 OR x=8) THEN
105                   EXIT loop_x;
106              END IF;
107         END LOOP loop_x;
108    END PROCESS;
109    END behavior;
```

Figure 7.11 Behavioral description of the coprocessor

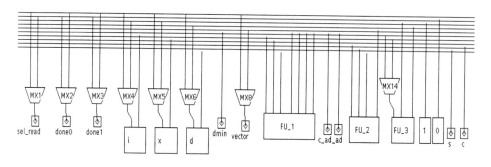

Figure 7.12 Datapath generation of *co_processor* by AMICAL

Figure 7.12 shows the datapath of the coprocessor, generated automatically by AMICAL from the VHDL description in figure 7.11. It schemes a MUX based architecture of the datapath. The FU_1, FU_2 and FU_3 correspond respectively to the functional units *min_addr*, *alu* and *incr*. The produced controller is a 7 state and 18 transition FSM. It controls the datapath through 23 control lines.

7.4.2 The mem_sequence

The subsystem *mem_sequence* is in charge of managing the sequence. It is connected to the coprocessor, to the top controller and to the external world as shown in figure 7.13.

The *mem_sequence* subsystem makes use of a memory block which is abstracted as a functional unit executing one procedure: write_ram. This procedure is used for loading the *sequence*. This memory is a dual port RAM where the first port

Case Study: Modular Design Using Behavioral Synthesis 231

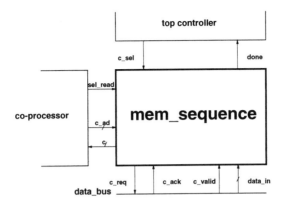

Figure 7.13 Connection of the *mem_sequence*

is used to connect the memory with the external world. The second port is used for communication with the coprocessor.

The communication with the coprocessor is not included in the main process description of the *mem_sequence*. The connections between the memory modules and the coprocessor are made in the architecture part using a port map statement. This communication will not be synthesized by AMICAL. It will be inserted during the personalization step. (e.g. section 5.4.3).

For the high-level synthesis of *mem_sequence* two functional units are needed: the RAM (dpram_8x8 for the procedure write_ram) and an incrementor (c_inc for the operation +).

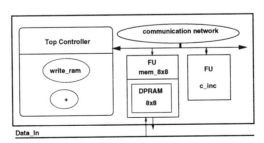

Figure 7.14 Abstract architecture of the *mem_sequence*

Figure 7.14 shows the abstract architecture of the *mem_sequence*. The procedure *write_ram* and the operator *+* are executed respectively by the functional units called *mem_8x8* and *c_inc*.

The behavioral description of *mem_sequence* is presented in figure 7.16, and commented in this section. *mem_sequence* is composed of a RAM, to store the *sequence*, and of a local controller.

The VHDL architecture consists of a behavioral part (process [l.32-66]) and a structural part [l.68-69]. The memorization component mem_8x8 is declared in the package pkg_components and instanced as sequence_mem in the structural part. The behavioral part consists of a unique process with three main steps: initializations [l.33-40], request [l.41-45] and loading [l.46-64].

During the initialization step the signals and variables are put to their initial values followed by a wait [l.38] until the validation of the activation signal: *c_sel*. The address and the output signals (such as the write enable signal *sel_write* of the RAM) are put to the initialization values.

The request step manages a request (c_req)/acknowledge (c_ack) protocol with external operators thanks to wait statements [l.43].

The writing step is the core of the process. It consists of a loop in which data_in is written in the sequence_ram at the incremented address a1 [l.56], as long as the end of the burst is not reached [l.47]. Two conditions need to be satisfied for writing, according to the chronogram illustrated by figure 7.15. First, the data arriving must be validated by the c_valid signal [l.47, l.54], one cycle before. Second, the write enable signal must be activated only when a data is validated [l.57] else the writing is interrupted [l.63].

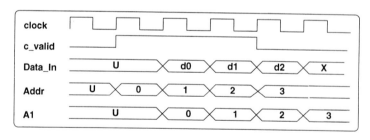

Figure 7.15 Writing in the RAM

The last step is the activation of the output control signal c_done signaling the end of the loading until the next selection [l.34].

The constraint of this description is to have a cycle-based behavior to fill the RAM. If c_valid equals to '1' for the whole burst without interruption, the

while loop is then executed each clock cycle. One wait statement appears in
the loop, in this case, for signal assignments. As the address (a1) must be
ready at the arrival of the data (data_in), its computation must be anticipated.
Consequently another signal (*addr*) is needed. The effective address assignment
is inserted inside a procedure declaration (in *write_ram*) [1.29] (figure 7.15). It
allows to respect the set-up time during the behavioral simulation (*after 2 ns*),
and to manipulate a pure behavioral description.

```
1   LIBRARY ieee;
2     USE ieee.std_logic_1164.ALL;
3     USE ieee.std_logic_unsigned.ALL;
4     USE work.pkg_types.ALL;
5     USE work.pkg_components.ALL;
6
7   ENTITY mem_sequence IS
8   PORT(  clk      : IN    bit1;
9          reset    : IN    bit1;
10         sel_read : IN    bit1;
11         c_sel    : IN    bit1;
12         c_req    : OUT   bit1;
13         c_ack    : IN    bit1;
14         c_valid  : IN    bit1;
15         data_in  : IN    bit8;
16         c_ad     : IN    bit3;
17         c        : OUT   bit8;
18         c_done   : OUT   bit1);
19  END mem_sequence;
20
21  ARCHITECTURE behavior OF mem_sequence IS
22  SIGNAL a1, addr         : bit3_r;
23  SIGNAL sel_write        : bit1;
24  BEGIN
25    PROCESS
26
27    PROCEDURE write_ram(sel_write: IN bit1; data: IN bit8) IS
28      BEGIN
29         a1 <= addr AFTER 2 ns;
30      END write_ram;
31
32    BEGIN
33      c_req <= '0';
34      c_done <= '1';
35      addr <= "000";
36      sel_write <= '0';
37
38      WAIT UNTIL (c_sel='1' AND rising_edge(clk) AND reset='1');
39
40      c_done <= '0';
41      c_req <= '1';
42
43      WAIT UNTIL (c_ack='1' AND rising_edge(clk)) OR reset/='1';
44
45      c_req <= '0';
```

```
46      write_loop: LOOP
47        IF c_valid/='1' AND addr<7 THEN
48
49          WAIT UNTIL (c_valid='1' AND rising_edge(clk)) OR reset/='1';
50
51          IF reset/='1' THEN
52            EXIT write_loop;
53          END IF;
54        END IF;
55        addr <= addr + 1;
56        write_ram(sel_write, data_in);
57        sel_write <= '1';
58        IF addr=7 OR reset/='1' THEN
59          WAIT UNTIL rising_edge(clk);
60          EXIT write_loop;
61        END IF;
62        WAIT UNTIL rising_edge(clk);
63        sel_write <= '0';
64      END LOOP write_loop;
65
66    END PROCESS;
67
68    sequence_mem : mem_8x8
69    PORT MAP ( c, data_in, c_ad, a1, sel_read, sel_write, clk);
70    END behavior;
```

Figure 7.16 behavioral description of *mem_sequence*

Figure 7.17 shows the datapath of the *mem_sequence*, generated automatically by AMICAL from the VHDL description in figure 7.16. It schemes a MUX based architecture of the datapath. The FU_1 and FU_2 correspond respectively to the mem_8x8 component and to the incrementor "c_inc". The produced controller is a 6 state and 15 transition FSM. It controls the datapath through 10 control lines.

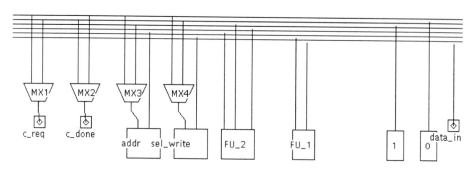

Figure 7.17 Datapath generation of *mem_sequence* by AMICAL

7.4.3 The mem_string

Figure 7.18 is presented the VHDL behavioral description of the module *mem_string*. It is the same as the description of *mem_sequence* except the memory size and the consideration of the burst number [l.30-31] in the *init* procedure declaration and the condition [l.55] signaling the end of the different bursts.

```
1   LIBRARY ieee;
2       USE ieee.std_logic_1164.ALL;
3       USE ieee.std_logic_unsigned.ALL;
4       USE work.pkg_types.ALL;
5       USE work.pkg_components.ALL;
6
7   ENTITY mem_string IS
8   PORT (  clk      : IN    bit1;
9           reset    : IN    bit1;
10          sel_read : IN    bit1;
11          s_sel    : IN    bit1;
12          burst    : IN    bit2_r;
13          s_req    : OUT   bit1;
14          s_ack    : IN    bit1;
15          s_valid  : IN    bit1;
16          data_in  : IN    bit8;
17          s_ad     : IN    bit5;
18          s        : OUT   bit8;
19          s_done   : OUT   bit1);
20  END mem_string;
21
22  ARCHITECTURE behavior OF mem_string IS
23  SIGNAL a1, addr      : bit5_r;
24  SIGNAL sel_write     : bit1;
25  BEGIN
26    PROCESS
27
28      PROCEDURE init(burst: IN bit2_r; SIGNAL a1: OUT bit5_r) IS
29      BEGIN
30        addr <= burst & "000";
31        a1   <= burst & "000";
32      END init;
33
34      PROCEDURE write_ram(sel_write: IN bit1; data: IN bit8) IS
35      BEGIN
36        a1 <= addr;
37      END write_ram;
38
39    BEGIN
40      s_req <= '0';
41      s_done <= '1';
42      addr <= "00000";
43      sel_write <= '0';
44
45      WAIT UNTIL (s_sel='1' AND rising_edge(clk) AND reset='1');
```

```
46
47      init(burst, addr);
48      s_done <= '0';
49      s_req <= '1';
50
51      WAIT UNTIL ((s_ack='1' AND rising_edge(clk)) OR reset/='1');
52
53      s_req <= '0';
54      write_loop: LOOP
55          IF s_valid/='1' AND addr/=7 AND addr/=15 AND addr<23 THEN
56
57                  WAIT UNTIL (s_valid='1' AND rising_edge(clk))
                                OR reset/='1';
58
59                  IF reset/='1' THEN
60                          EXIT write_loop;
61                  END IF;
62          END IF;
63          addr <= addr + 1;
64          write_ram(sel_write, data_in);
65          sel_write <= '1';
66          IF addr=7 OR addr=15 OR addr>=23 OR reset/='1' THEN
67                  WAIT UNTIL rising_edge(clk);
68                  EXIT write_loop;
69          END IF;
70          WAIT UNTIL rising_edge(clk);
71          sel_write <= '0';
72      END LOOP write_loop;
73
74      END PROCESS;
75
76      string_mem : mem_24x8
77      PORT MAP (s, data_in, s_ad, a1, sel_read, sel_write, clk);
78
79      END behavior;
```

Figure 7.18 Behavioral description of *mem_string*

The abstract architecture of *mem_string* is the same as the abstract architecture of *mem_sequence*. The extra procedure *init* is executed by the functional unit *s_inc*.

Figure 7.19 shows the datapath of the *mem_string*, generated automatically by AMICAL from the VHDL description in figure 7.18. It schemes a MUX based architecture of the datapath. The FU_1 and FU_2 correspond respectively to the mem_24x8 component and to the incrementor "s_inc". The produced controller is a 6 state and 15 transition FSM. It controls the datapath through 11 control lines.

Case Study: Modular Design Using Behavioral Synthesis

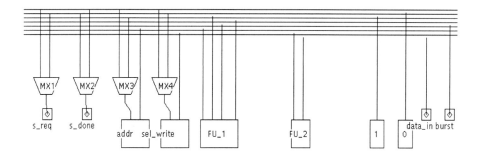

Figure 7.19 Datapath generation of *mem_string* by AMICAL

7.5 SYSTEM DESIGN

This section describes the high-level synthesis of the global *WSS*. For the validation, four behavioral descriptions must be available: the two memory modules, the coprocessor and the top controller. After the synthesis of the three subsystems, the synthesis of the controller can be performed.

The function of the WSS was already described in this chapter. We will use the algorithm described in figure 7.3(b) and the partitioning represented by the abstract architecture of figure 7.4.

The WSS interacts directly with the external world through two signals: command (starts computation) and status (indicates a busy state or the availability of a result). The interface of the WSS includes several other signals that are used by the coprocessor and the memory modules to communicate with the external world. The interconnection scheme is given in figure 7.20.

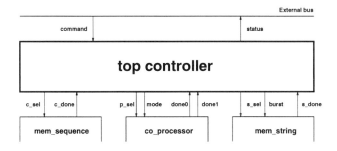

Figure 7.20 Connections of the WSS

The communication between the top Controller and the subsystems makes use of a quite simple protocol (figure 7.21). In order to activate a component, the top controller needs to set the corresponding select signals. When a component is busy, it resets the corresponding done signal. This interconnection of the top control and the subsystem is shown in figure 7.20.

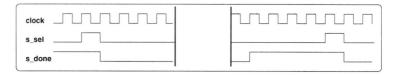

Figure 7.21 Activation protocol of *mem_string*

As explained in chapter 4 there are two ways for describing such a system. The subsystems may be used as external components or as functional units (figures 7.4). In the first case, the communication signals are directly accessible in the behavioral description. Figure 7.22 gives a VHDL description of the top controller as separate unit. The whole WS system is described in figure 7.23 as an assembly of four components: top, *mem_sequence*, *mem_string* and *co_processor*. The VHDL architecture of *top* is composed of a process that models the top controller. In *WSS* the structural architecture contains this *top* and three instances corresponding to the three subsystems. In this case, any change in the communication protocol between the components and the top controller would imply a modification of the behavioral description of the top controller. For instance, *mem_string* is activated three times within the top controller. If we change the communication protocol between the top controller and *mem_string*, we will have to change the description of the top controller at three different locations (figure 7.22 [1.51, 64, 77]).

```
 1  LIBRARY ieee;
 2    USE ieee.std_logic_1164.ALL;
 3    USE ieee.std_logic_arith.ALL;
 4    USE ieee.std_logic_unsigned.ALL;
 5    USE work.pkg_types.ALL;
 6
 7  ENTITY top IS
 8  PORT (   clk               : IN     bit1;
 9           reset             : IN     bit1;
10           command           : IN     bit1;
11           c_sel             : OUT    bit1;
12           s_sel             : OUT    bit1;
13           burst             : OUT    bit2_r;
14           p_sel             : OUT    bit1;
15           mode              : OUT    bit1;
16           c_done            : IN     bit1;
```

```
17            s_done            : IN    bit1;
18            done0             : IN    bit1;
19            done1             : IN    bit1;
20            status            : OUT   bit1);
21  END top;
22
23  ARCHITECTURE behavior OF top IS
24  CONSTANT n              : integer:= 2;
25  SIGNAL    burst_int     : bit2_r;
26  BEGIN
27    PROCESS
28
29    BEGIN
30      mode <='0';
31      burst_int <= "00";
32      p_sel <= '0';
33      c_sel <= '0';
34      s_sel <= '0';
35      status <= '0';
36      main_loop : LOOP
37
38         WAIT UNTIL c_done ='1' AND s_done ='1' AND
                      done0='1' AND done1='1' AND
39                    rising_edge(clk) AND reset='1';
40
41         status <= '1';
42         IF command/='1' THEN
43
44            WAIT UNTIL command='1' AND reset='1' AND rising_edge(clk);
45
46         END IF;
47         mode <='0';
48         burst_int <= "00";
49         status  <= '0';
50         c_sel <= '1';
51         s_sel <= '1';
52
53         WAIT UNTIL c_done='0' AND s_done = '0' AND rising_edge(clk);
54
55         c_sel <= '0';
56         s_sel <= '0';
57
58         WAIT UNTIL c_done='1' AND s_done = '1' AND rising_edge(clk);
59
60         IF reset/='1' THEN
61             EXIT main_loop;
62         END IF;
63         burst_int <= burst_int + 1;
64         s_sel <= '1';
65
66         WAIT UNTIL s_done = '0' AND rising_edge(clk);
67
68         s_sel <= '0';
69
70         WAIT UNTIL s_done = '1' AND rising_edge(clk);
71
```

```
72       IF reset/='1' THEN
73          EXIT main_loop;
74       END IF;
75       WHILE burst_int < n AND reset='1' LOOP
76          burst_int <= burst_int + 1;
77          s_sel <= '1';
78          p_sel <= '1';
79
80          WAIT UNTIL s_done='0' AND done0='0' AND rising_edge(clk);
81
82          s_sel <= '0';
83          p_sel <= '0';
84          WAIT UNTIL s_done='1' AND done0='1' AND rising_edge(clk);
85
86       END LOOP;
87       IF reset/='1' THEN
88          EXIT main_loop;
89       END IF;
90       mode <= '1';
91       p_sel <= '1';
92
93       WAIT UNTIL done1='0' AND rising_edge(clk);
94
95     END LOOP main_loop;
96   END PROCESS;
97
98   burst <= burst_int;
99
100 END behavior;
```

Figure 7.22 VHDL behavioral description of the top controller instanced within a structural architecture of the WSS

```
1   LIBRARY ieee;
2     USE ieee.std_logic_1164.ALL;
3     USE ieee.std_logic_arith.ALL;
4     USE ieee.std_logic_unsigned.ALL;
5     USE work.pkg_types.ALL;
6     USE work.pkg_components.ALL;
7
8   ENTITY wss IS
9   PORT ( reset           : IN    bit1;
10         clk             : IN    bit1;
11         command         : IN    bit1;
12         c_req           : OUT   bit1;
13         c_ack           : IN    bit1;
14         c_valid         : IN    bit1;
15         s_req           : OUT   bit1;
16         s_ack           : IN    bit1;
17         s_valid         : IN    bit1;
18         data_in         : IN    bit8;
19         dmin            : OUT   bit11;
20         vector          : OUT   bit4;
21         status          : OUT   bit1);
22  END wss;
```

Case Study: Modular Design Using Behavioral Synthesis

```
23
24  ARCHITECTURE rtl OF wss IS
25    CONSTANT n                      : integer:= 2;
26    SIGNAL    c_sel, c_done         : bit1;
27    SIGNAL    s_sel, s_done         : bit1;
28    SIGNAL    p_sel, done0, done1   : bit1;
29    SIGNAL    mode, sel_read        : bit1;
30    SIGNAL    burst                 : bit2_r;
31    SIGNAL    s_ad                  : bit5;
32    SIGNAL    c_ad                  : bit3;
33    SIGNAL    s, c                  : bit8;
34  BEGIN
35
36    tp: top
37    PORT MAP(clk, reset, command, c_sel, s_sel, burst, p_sel, mode,
                c_done, s_done, done0, done1, status);
38
39    mq: mem_sequence
40    PORT MAP(clk, reset, sel_read, c_sel, c_req, c_ack,
                c_valid, data_in, c_ad, c, c_done);
41
42    mt: mem_string
43    PORT MAP(clk, reset, sel_read, s_sel, burst, s_req, s_ack,
                s_valid, data_in, s_ad, s, s_done);
44
45    pr: co_processor
46    PORT MAP(clk, reset, c, s, p_sel, mode, sel_read, c_ad, s_ad,
                dmin, vector, done0, done1);
47
48  END rtl;
```

Figure 7.23 VHDL structural description of the *WSS*

Another solution for describing this model is to hide the communication protocol between the top controller and the subsystems using procedures. A new description of the WSS is given in figure 7.24. The selection of *mem_string* is made by one procedure called *write_string*. In this case a change of the selection protocol of *mem_string* will induce only a single change in the behavioral description of the top controller. The procedure "write_string" [l.42-45] which is in charge of selecting *mem_string* is the only part of the code that has to be changed. Of course when the communication protocol gets more complex and makes use of several signals, this hiding of protocols becomes much more attractive.

```
1  LIBRARY ieee;
2    USE ieee.std_logic_1164.ALL;
3    USE ieee.std_logic_arith.ALL;
4    USE ieee.std_logic_unsigned.ALL;
5    USE work.pkg_types.ALL;
6    USE work.pkg_components.ALL;
7
```

```
8    ENTITY wss IS
9    PORT ( reset              : IN    bit1;
10          clk                : IN    bit1;
11          command            : IN    bit1;
12          c_req              : OUT   bit1;
13          c_ack              : IN    bit1;
14          c_valid            : IN    bit1;
15          s_req              : OUT   bit1;
16          s_ack              : IN    bit1;
17          s_valid            : IN    bit1;
18          data_in            : IN    bit8;
19          dmin               : OUT   bit11;
20          vector             : OUT   bit4;
21          status             : OUT   bit1);
22   END wss;
23
24   ARCHITECTURE behavior OF wss IS
25    CONSTANT n                       : integer:= 2;
26    SIGNAL   c_sel, c_done            : bit1;
27    SIGNAL   s_sel, s_done            : bit1;
28    SIGNAL   p_sel, done0, done1      : bit1;
29    SIGNAL   mode, sel_read           : bit1;
30    SIGNAL   burst                    : bit2_r;
31    SIGNAL   s_ad                     : bit5;
32    SIGNAL   c_ad                     : bit3;
33    SIGNAL   s, c                     : bit8;
34   BEGIN
35    PROCESS
36
37     PROCEDURE write_sequence(val : IN bit1) IS
38     BEGIN
39       c_sel <= val;
40     END write_sequence;
41
42     PROCEDURE write_string(val : IN bit1; burst: IN bit2_r) IS
43     BEGIN
44       s_sel <= val;
45     END write_string;
46
47     PROCEDURE run(val : IN bit1) IS
48     BEGIN
49       mode <= val;
50       p_sel <= '1' AFTER 2 ns, '0' AFTER 22 ns;
51     END run;
52
53    BEGIN
54
55     WAIT UNTIL done0 = '1' AND done1 = '1' AND rising_edge(clk);
56
57     burst <= "00";
58     write_sequence('0');
59     write_string('0', "00");
60     status <= '1';
61     IF command/='1' THEN
62
63       WAIT UNTIL command='1' AND rising_edge(clk);
```

```
64
65    END IF;
66    status    <= '0';
67    write_sequence('1');
68    write_string('1', burst);
69
70    WAIT UNTIL c_done='0' AND s_done = '0' AND rising_edge(clk);
71
72    write_sequence('0');
73    write_string('0', burst);
74
75    WAIT UNTIL c_done='1' AND s_done = '1' AND rising_edge(clk);
76
77    burst <= burst + 1;
78    write_string('1', burst);
79
80    WAIT UNTIL s_done = '0' AND rising_edge(clk);
81
82    write_string('0', burst);
83
84    WAIT UNTIL s_done = '1' AND rising_edge(clk);
85
86    WHILE burst < n LOOP
87      burst <= burst + 1;
88      write_string('1', burst);
89
90      WAIT UNTIL s_done = '0' AND rising_edge(clk) ;
91
92      write_string('0', burst);
93      run('0');
94
95      WAIT UNTIL s_done='1' AND done0='1' AND rising_edge(clk);
96
97    END LOOP;
98    run('1');
99
100   END PROCESS;
101
102   mq: mem_sequence
103   PORT MAP(clk, reset, sel_read, c_sel, c_req, c_ack,
                   c_valid, data_in, c_ad, c, c_done);
104
105   mt: mem_string
106   PORT MAP(clk, reset, sel_read, s_sel, burst, s_req, s_ack,
                   s_valid, data_in, s_ad, s, s_done);    107
108   pr: co_processor
109   PORT MAP(clk, reset, c, s, p_sel, mode, sel_read, c_ad, s_ad,
                   dmin, vector, done0, done1);
110
111   END behavior;
```

Figure 7.24 VHDL behavioral description of the WSS

Two descriptions given in figures 7.22 and 7.24 imply two different synthesis schemes, when using AMICAL:

- The subsystems are ignored by behavioral synthesis (description figure 7.22). The procedures are inlined within the description of the top controller.

- The subsystems are handled as functional units (description figure 7.24). In this case the procedures will be interpreted as operations executed on coprocessors (components). In addition to the behavioral VHDL, we need to abstract the subsystems as functional units in order to reuse them.

Figure 7.25 shows the datapath of the *WSS*, generated automatically by AMICAL from the VHDL description in figure 7.24. It schemes a MUX based architecture of the datapath. The FU_1, FU_2, FU_3, and FU_4 correspond respectively to *mem_sequence*, *mem_string*, *co_processor* and to the incrementor "w_inc". The produced controller is a 9 state and 21 transition FSM. It controls the datapath through 9 control lines.

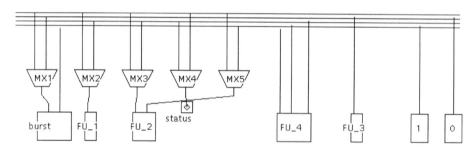

Figure 7.25 Datapath generation of *WSS* by AMICAL

7.6 BEHAVIORAL AND RTL SIMULATIONS

For such a system design, the verification needs mixed level simulations validating the high level synthesis of each subsystem and finally of the global system. Figure 7.26 shows the different simulations for each component inside the global system.

Simulation 1 validates the behavioral descriptions of the *WSS* and of the component, i.e. the global functionality.

Simulation 2 validates the high-level synthesis of the component, but is not necessary, as the protocol modeling is the same as in the first simulation.

Simulation 3 validates the high-level synthesis of the *WSS* and so, of the protocol abstraction corresponding to the component.

Simulation 4 validates both RTL descriptions before the logic and RTL synthesis.

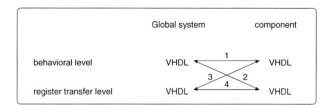

Figure 7.26 Mixed level simulations

7.7 SUMMARY

This chapter discussed a modular application of behavioral synthesis for the design of a window searching system.

The system is decomposed into four modules: a coprocessor, two memory modules and a top controller. The design process starts with the synthesis of the three first components. The top controller may be designed as a simple component interacting with three other modules or as a processor that makes use of the three first modules as behavioral components.

For each module we detailed the function, the abstract architecture, the VHDL description and the architecture produced by AMICAL. The VHDL description style mixes cycle fixed I/O mode and behavioral state fixed I/O mode. It was constrained in order to allow the simulation of the behavioral model and the RTL model produced by AMICAL using the same testbench.

REFERENCES

[AC91]   I. Ahmad and C. Y. Roger Chen. Post-processor for data path synthesis using multiport memories. In *Proceedings of the International Conference on Computer-Aided Design*, pages 276–279, November 1991.

[Anc86]   François Anceau. *The Architecture of Microprocessors*. Addison-Westley, 1986.

[AP89]   P. Anderson and L. Philpson. MOVIE - an Interactive Environment for Silicon compilation tools. *IEEE trans. on CAD*, 6, June 1989.

[BCP91]   R.A. Bergamaschi, R. Camposano, and M. Payer. Area and performance optimizations in path-based scheduling. In *Proceedings of the European Conference on Design Automation*, pages 450–455, Feb 1991.

[BDB94]   S. Bhatacharya, S. Dey, and F. Brglez. Performance analysis and optimization of schedules for conditional and loop intensive specifications. In *Proceedings of the Design Automation Conference*, pages 491–496, 1994.

[BK+92]   J. Biesenack, M. Koster, et al. The Siemens High-Level Synthesis System, CALLAS. In *Intl. High-Level Synthesis Wkshop*, November 1992.

[Cam91]   R. Camposano. Path-based scheduling for synthesis. *IEEE trans. on CAD*, 10(1):85–93, January 1991.

[CB90]   R. Camposano and R.A. Bergamaschi. Synthesis using path-based scheduling: Algorithms and exercises. In *Proceedings of the Design Automation Conference*, pages 450–455, June 1990.

[CBH+91]   R. Camposano, R.A. Bergamaschi, C. E. Haynes, M. payer, and S. M. Wu. *High Level VLSI Synthesis*, chapter The IBM High-Level Synthesis System. Kluwer Academic Publishers, 1991.

[Cle88]   J. C. Cleaveland. Building application generators. *IEEE trans. on Software*, July 88.

[CLS93]   L.-F. Chao, A. LaPaugh, and E. H.-M. Sha. Rotation scheduling: A loop pipelining algorithm. In *Proceedings of the Design Automation Conference*, pages 566–562, June 1993.

[CPTR89]   C. Chu, M. Pontkonjak, M. Thaler, and J. Rabaey. HYPER: An Interactive Synthesis Environment for High Performance Real Time Applications. In *Proceeding ICCD'89*, pages 432–435, Massachusetts, 1989.

[CST91]   R. Camposano, L.F. Saunders, and R.M. Tabet. VHDL as Input for High-Level Synthesis. *IEEE Design and Test of Computers*, pages 43–49, March 1991.

[CT89]   R. Camposano and R.M. Tablet. Design representation for the synthesis of behavioural VHDL models. In *Proceedings of CHDL*, May 1989.

[DiL88]   C. DiLisi. Computers in Molecular Biology: Current Applications and Emerging Trends. *Science*, pages 47–57, April 1988. number 240.

[DPA91]   D.J.Kinniment, P.P.Acarnley, and A.G.Jack. An integrated circuit controller for brushless dc drives. In *European Power Electronics Conference Proceedings*, volume 4, pages 111–116, EPE Firenze, 1991.

[DR85]   Peter Denyer and David Renshaw. *VLSI SIGNAL PROCESSING: A BIT-SERIAL APPROACH*. Addison-Wesley Publishing Compagny, 1985.

[FF95]   W. B. Frakes and C. J. Fox. Sixteen questions about software reuse. *Communications of the ACM*, 38(6):75–87, June 1995.

[FI94]   W. B. Frakes and S. Isoda. Success factors and systematic reuse. *IEEE trans. on Software*, September 1994.

[GC93]   E.F. Girczyc and S. Carlson. Increasing design quality and engineering productivity through design re-use. In *Proceedings of the Design Automation Conference*, June 1993.

[GDWL92]   D. Gajski, N. Dutt, A. Wu, and Y. Lin. *High-Level Synthesis : Introduction to Chip and System Design*. Kluwer Academic Publishers, Boston, Massachusetts, 1992.

References

[GE92]   C. H. Gebotys and M. I. Elmasry. *Optimal VLSI Architectural Synthesis: Area, Performance, Testability*. Kluwer Academic Publishers, 1992.

[GKP85]   J. Granacki, D. Knapp, and A.C. Parker. The ADAM Advanced Design Automation System: Overview, Planner and Natural Language Interface. In *Proceedings of the Design Automation Conference*, June 1985.

[GR94]   Daniel D. Gajski and L. Ramachandran. Introduction to high-level synthesis. *IEEE Design and Test of Computers*, pages 44–54, Winter 1994.

[GVM89]   G. Goossens, J. Vandewalle, and H. De Man. Loop optimization in register-transfer scheduling for DSPsystems. In *Proceedings of the Design Automation Conference*, pages 826–831, June 1989.

[GW92]   D.D Gajski and W. Wolf. *High-Level Synthesis*. Kluwer Academic Publishers, 1992.

[GW95]   G.Schumacher and W.Nebel. Inheritance concept for signals in object-oriented extensions to vhdl. In *Proceedings of the European Design Automation Conference*, 1995.

[H$^+$90] D. Hare et al. Statecharts: A working environment for the development of complex reactive systems. *IEEE trans. on Software Engineering*, 16(4), April 90.

[HH93]   C. T. Hwang and Y. C. Hsu. Zone scheduling. *IEEE trans. on CAD*, 12(7), July 1993.

[Hil85]   P.N Hilfinger. A high level language and silicon compiler for digital signal processing. *IEEE Custom Integrated Circuits Conference*, pages 213–216, 1985.

[HT83]   C. Y. Hitchcock and D. E. Thomas. A method of automatic data path synthesis. In *Proceedings of the Design Automation Conference*, 1983.

[Hu61]   T. C. Hu. Parallel sequencing and assembly line problems. *Operations Research*, pages 841–848, November 1961.

[Inc94]   Synopsys Inc. *Synopsys Behavioral Compiler User Guide, Version 3.2a*, October 1994.

[JCG94] H. Juan, V. Chaiyakul, and D.D. Gajski. Condition graphs for high-quality behavioral synthesis. In *Proceedings of the International Conference on Computer-Aided Design*, pages 170–174, November 1994.

[Jon93] G.G Jong. *Generalized Data Flow Graphs, Theory and Applications*. PhD thesis, Eindhoven University of Technology, 1993.

[JPO93] A.A. Jerraya, I. Park, and K. O'Brien. AMICAL: An Interactive High Level Synthesis Environment. In *Proceedings of the European Conference on Design Automation*, February 1993.

[KD92] D.C. Ku and G. DeMicheli. Relative scheduling under timing constraints. *IEEE trans. on CAD*, May 1992.

[KGM95] A. Kifli, G. Goossens, and H. De Man. A unified scheduling model for high-level synthesis and code generation. In *Proceedings of the European Conference on Design Automation*, Paris, France, Mars 1995.

[Kis96] Polen Kission. *Exploitation de la hiérarchie et de la ré-utilisation de blocs existants par la synthèse de haut niveau*. PhD thesis, INPG, 1996.

[KL93] T. Kim and C. L. Liu. Utilization of multiport memories in data path synthesis. In *Proceedings of the Design Automation Conference*, pages 298–302, 1993.

[KLMM95] D. Knapp, T. Ly, D. MacMillen, and R. Miller. Behavioral synthesis methodology for HDL-based specification and validation. In *Proceedings of the Design Automation Conference*, pages 286–291, June 1995.

[Kna96] D. Knapp. *Behavioral Synthesis*. Prentice Hall, 1996.

[KNRR88] H. Krämer, M. Neher, G. Rietsche, and W. Rosenstiel. Data path and control synthesis in the CADDY system. In *International Workshop at INPG*, Grenoble, 1988. INPG.

[KP87] F. Kurdahi and A. C. Parker. REAL: A program for Register Allocation. In *Proceedings of the Design Automation Conference*, pages 210–215, June 1987.

[LCGM93] D. Lanneer, M. Cornero, G. Goossens, and H. De Man. An assignment technique for incompletely specified data-paths. In *Proceedings of the European Conference on Design Automation*, Paris, February 1993.

References

[LG88]   J.S. Lis and D.D. Gajski. Synthesis from VHDL. In *Proceedings of the International Conference on Computer-Aided Design*, pages 378–381, October 1988.

[LHCF96] M. T-C. Lee, Y-C. Hsu, B. Chen, and M. Fujita. Domain-Specific High-Level Modelling and Synthesis for ATM Switch Design Using VHDL. In *Proceedings of the Design Automation Conference*, pages 585–590, 1996.

[LMWV91] P.E.R. Lippens, J.L.V. Meerbergen, A. V. Werf, and W.F.J. Verhaegh. PHIDEO, A silicon Compiler for High Speed Algorithms. In *Proceedings of the European Conference on Design Automation*, pages 436–441, Amsterdam, March 1991.

[MC80]   C. A. Mead and L. A. Conway. *Introduction to VLSI Sytems*. Addison-Wesley Publishing Company, 1980.

[MCG+90] H. De Man, F. Catthoor, G. Goossens, J. Van Meerbergen, S. Note, and J. Huisken. Architecture-driven synthesis techniques for VLSI implementation of DSP algorithms. *Proc. IEEE*, 78(2):319–355, February 1990.

[Mic94]   G. De Micheli. *Synthesis and Optimization of Digital Circuits*. Mc Graw Hill, 1994.

[MK88]   G. De Micheli and D.C. Ku. HERCULES - a system for high-level synthesis. In *Proceedings of the Design Automation Conference*, 1988.

[MLD92]   P. Michel, U. Lauther, and P. Duzy. *The Synthesis Approach to Digitual System Design*. Kluwer Academic Publishers, 1992.

[MMM95]   H. Mili, F. Mili, and A. Mili. Re-using software issues and research directions. *IEEE trans. on Software*, 21(6), June 95.

[MPC90]   M.C. McFarland, A.C. Parker, and R. Camposano. The high-level synthesis of digital systems. *IEEE*, 78(2):301–318, February 1990.

[MSDP93] E. Martin, O. Sentieys, H. Dubois, and J. L. Philippe. GAUT: An architectural synthesis tool for dedicated signal processors. In *Proceedings of the European Design Automation Conference*, 1993.

[N+91]   S. Note et al. CATHEDRAL III: Architecture-driven High-Level Synthesis for High Throughput DSP Applications. In *Proceedings of the Design Automation Conference*, 1991.

[NBD92]   V. Nagasamy, N. Berry, and C. Dangelo. Specification, Planning, and Synthesis in a VHDL Design Environment. *IEEE Design and Test of Computers*, pages 58–68, June 1992.

[NSM89]   J. Nestor, B. Soudan, and Z. Mayet. Mies a micro-architecture design tool. In *Proc. 22nd International Workshop on Microprogramming and Micro-Architecture*, 1989.

[OG86]   A. Orailogo and D.D Gajski. Flow graph representation. In *Proceedings of the Design Automation Conference*, June 1986.

[OM96]   K. O'Brien and S. Maginot. Towards Maximising the use of VHDL for Synthesis. In *Proceedings of the European Design Automation Conference*, Geneve, SW, September 1996.

[ORJ93]   K. O'Brien, M. Rahmouni, and A.A. Jerraya. DLS: A scheduling algorithm for high-level synthesis in VHDL. In *Proceedings of the European Conference on Design Automation*, Paris, France, February 1993.

[Pan88]   B.M. Pangrle. SPLICER: A Heuristic Approach to Connectivity Binding. In *Proceedings of the Design Automation Conference*, 1988.

[Pen86]   Z. Peng. Synthesis of VLSI systems with the CAMAD design aid. In *Proceedings of the Design Automation Conference*, 1986.

[PK89]   P.G. Paulin and J.P. Knight. Force-directed scheduling for the behavioral synthesis of ASIC's. *IEEE trans. on CAD*, 8(6):661–679, June 1989.

[PKG86]   P.G. Paulin, J.P. Knight, and E.F. Girczyc. HAL: a multi-paradigm approach to automatic data path synthesis. In *Proceedings of the Design Automation Conference*, 1986.

[RBL+96] M. Rahmouni, M. BenMohammed, C. Liem, H. Ding, P. Kission, and A.A. Jerraya. The Applications of Synthesis Techniques for the Generation of Instruction-Set Architectures. *Integrated Computer-Aided Engeneering*, 1996. Submitted.

[RGC94]   L. Ramachandran, D.D. Gajski, and V. Chaiyakul. An Algorithm for Array Variable Clustering. In *Proceedings of the European Conference on Design Automation*, Paris, France, 1994.

References

[RJ95a] M. Rahmouni and A.A. Jerraya. Formulation and Evaluation of Scheduling Techniques for Control Flow Graphs. In *Proceedings of the European Design Automation Conference*, Brighton, UK, September 1995.

[RJ95b] M. Rahmouni and A.A. Jerraya. PPS: A Pipeline Path based Scheduler. In *Proceedings of the European Conference on Design Automation*, Paris, France, March 1995.

[RJ96a] M. Rahmouni and A.A. Jerraya. A Taxonomy of Scheduling Algorithms for Control Flow Dominated Specifications. *IEEE trans. on CAD*, 1996. Submitted.

[RJ96b] M. Rahmouni and A.A. Jerraya. Performance Analysis based Scheduling for Programmable Architectures. Internal report, TIMA/INPG, 1996.

[ROJ94] M. Rahmouni, K. O'Brien, and A.A. Jerraya. A loop-based scheduling algorithm for hardware description languages. *Parallel Processing Letters*, 4(3):351–364, 1994.

[Sau87] L.F. Saunders. The IBM VHDL Design System. In *Proceedings of the Design Automation Conference*, pages 484–490, 1987.

[SDL87] Computer Networks and ISDN Systems. *CCITT SDL*, 1987.

[Seq83] C. H. Sequin. Managing VLSI complexity: an outlook. *Proceedings of the IEEE*, 71(1), January 1983.

[Seq86] C. H. Sequin. VLSI design stategies. In *Proceedings of the summer school on VLSI tools and applications*. Kluwer Academic Publisher, July 1986.

[SIA94] The National Technology Roadmap for SemiConductors. Technical report, SemiConductor Industry Association, San Jose, California, 1994.

[SMG95] S. Swanny, A. Molin, and B. Govnot. OO-VHDL: Object-oriented extensions to VHDL. *Computer*, October 1995.

[ST95] H. Schmit and D. Thomas. Array Mapping in Behavioral Synthesis. In *Intl. Sym. Sys. Synthesis*, September 1995.

[Sta94] W. Staringer. Constructing applications from reusable components. *IEEE trans. on Software*, September 1994.

[Sto91]   L. Stok. *Architectural Synthesis and Optimization of Digital Systems*. PhD thesis, Eindhoven University of Technology, 1991.

[TDW+88]   D.E. Thomas, E.M. Dirkes, R.A. Walker, J.V. Rajan, J.A. Nestor, and R.L. Blackburn. The system architect's workbench. In *Proceedings of the Design Automation Conference*, pages 337–343, June 1988.

[TJ95]   A.H. Timmer and J.A.G. Jess. Exact Scheduling Strategies based on Bipartie Graph Matching. In *Proceedings of the European Conference on Design Automation*, Paris, France, Mars 1995.

[TM91]   D. E. Thomas and P. Moorby. *The VERILOG Hardware Description Language*. Kluwer Academic Publishers, 1991.

[Tri85]   Howard Trickey. *Compiling Pascal Programs into Silicon*. PhD thesis, Department of Computer Science, Stanford University, July 1985.

[TRLG81]   S. Trimberger, J. A. Rowson, C. R. Lang, and J. P. Gray. A structered design methodology and associated software tools. *IECS*, 28(7), July 1981.

[VHD87]   IEEE, NY. USA. *IEEE Standard VHDL Langage Reference Manual*, March 1987.

[VRB+93]   J. Vanhoof, K. V. Rompaey, I. Bolsens, G. Goossens, and H. De Man. *High-Level Synthesis for Real-Time Digitial Signal Processing*. Kluwer Academic Publishers, 1993.

[WC95]   Robert A. Walker and S. Chaudhuri. Introduction to the scheduling problem. *IEEE Design and Test of Computers*, pages 60–68, Winter 1995.

[Wis89]   Adrian Wise. *Introduction To Motion Picture Coding and the CCITT Algorithm*, December 1989.

[Wol91]   W. Wolf. *High Level VHDL Synthesis*, chapter Architectural Optimization Methods For Control Dominated Machines. Kluwer Academic Publishers, 1991.

[WT87]   R.A. Walker and D. E. Thomas. Design representation and transformation in the system architect's workbench. In *Proceedings of the International Conference on Computer-Aided Design*, pages 166–169, 1987.

[Zim79]   G. Zimmermann. The MIMOLA Design System: A Computer Aided Digital Processor Design Method. In *Proceedings of the Design Automation Conference*, 1979.

INDEX

INDEX

A

Abstraction for reuse, **132**
Abstraction levels, **2**
AFAP, **110**
See As Fast As Possible scheduling
ALAP, **97**
See As Late As Possible scheduling
Allocation, **12**, 15–16
AMICAL, 17, 21, 33, **155**, 213, 217, 233, 247
  architectural exploration, **191**–194
  design loop, **194**–199
  design steps, 155–**156**, 161–162
    architecture generation, **183**–185
    connection allocation, **178**–182
    functional unit allocation, **176**
    input description, **162**–165
    macro-scheduling, **172**,–175
    micro-scheduling, **177**–178
    VHDL compilation, **171**
  incremental refinement model, **157**, 162
  interactive synthesis, **185**–190
  library of components, **161**, 166–167
    FU, **167**–171
  target architecture, **159**–160
  VHDL subset, **157**–159, 164
Architectural synthesis, **2**
See Behavioral synthesis
Architecture generation, **12**, 15

As Fast As Possible scheduling, **110**–113, 118
As Late As Possible scheduling, **97**
As Soon As Possible scheduling, **97**
ASAP, **97**
See As Soon As Possible Scheduling

B

Behavioral component, **137**–140, 147, 206–210, 227–240
  modeling for reuse
    mixed behavioral-structural model, **148**–149
    pure behavioral VHDL model, **148**–149
Behavioral description languages, **11**
  VHDL, 67–**68**, 68–97, 223–227
Behavioral level, 4, **70**–89
Behavioral state mode, **92**
Behavioral synthesis, 2–**5**, 23
  application domains, **7**–8
  control flow oriented, 8, **109**–118
  data flow oriented, 8, **96**–104
  mixed data control flow oriented, **8**–9
  case study, **203**–248
  internal representation or intermediate form, **9**, 24, 29
  steps, **12**–16
  target architecture, 9, 11, **33**
  tools, **6**–7, 11, 16

Binding, **12**, 15
Branch probability, **109**
Bus, **55**

C

CDFG, **24**
  See Control Data Flow Graph
CFBS, **105**
  See Scheduling, Control Flow Based
CFG, **24**
  See Control Flow Graph
Chaining, **50**
Clock cycle, **35**, 60
Communication topology
  point-to-point topology, **46**
  shared-interconnection topology, **46**
Communication unit, 35, **46**
Component
  abstraction, 20, 132–133, **137**–147, 209
  reuse, 17–20, **131**
  VHDL modeling, **144**–147
Connexion allocation, **15**
Constrained scheduling, **98**–104
Control Data Flow Graph, **27**–28
Control Flow Graph, **25**–27, 94, 105, 121
Control part, **33**
  See Controller
Controller, 12, 33, **58**–61, 72, 77, 118
  organization, **58**
    hardwired control model, **61**
    programmable control model, **63**
Coprocessor, 31–33, 83–85, 136, 140, 217, 220, 228–229, 233–234
Cost function, **108**–113
CU, **46**
  See Communication unit

Cycle-fixed mode, **91**

D

Data Flow Graph, **24**–25, 50, 94–96
Datapath controller model, 12, **33**
  structure, **33**
  synchronization, **35**
Datapath, 12, 33–34, **40**, 233
  functional organization, 40, **49**
  pipelining, **52**
  representation, **41**
  structure, **34**, 40
    components, **42**, 45–46, 48
  target architecture, **54**
    bus based architecture, 47, **55**
    multiplexer based architecture, 47, **56**
Design abstraction, **132**
Design reuse, **127**
  See Reuse
Design verification, **247**–248
DFBS, **95**
  See Scheduling, Data Flow Based
DFG, **24**
  See Data Flow Graph
DLS, **113**
  See Dynamic Loop Scheduling
Dynamic Loop scheduling, **113**–114, 118
ECU, **48**
  See External Communication Unit
Execution part, **33**
  See Datapath
Execution thread, **90**
External communication unit, 35, **48**
FDS, **100**
  See Force Directed scheduling
Finite-State Machine, **58**, 121
  Mealy, **59**–63, 108
  Moore, **59**–63

Index   261

sequencing, **60**
theory, **58**
Fixed I/O mode, **93**
Force Directed scheduling, **100**–102
Free-floating I/O mode, **93**
FSM, **58**
See Finite State Machine
FSMC, **31**
See Intermediate form
FSMD, **29**
See Intermediate form
FU, **42**
See Functional unit
Function calls, 82, **146**–149
Functional unit, 35, **42**, 137, 145–147
  organization, **42**
  pipelining, **43**

G

Gate level, **2**

H

Hierarchical architecture, **31**
Hierarchical decomposition, 19–20, **205**
Hierarchy, **18**
High-level synthesis, **2**
See Behavioral synthesis

I

ILP, **103**
See Integer Linear Programming Scheduling
Integer Linear Programming scheduling, **103**–105
Intermediate form, **10**, 12–13, **23**, 64
  architecture oriented, 10, **29**
    FSMC, 31–**32**, 161, 172–175
    FSMD, **29**

language oriented, 10, **24**
  CDFG, **24**, 27
  CFG, **24**–25
  DFG, **24**–25

L

Latency, **44**, 53
Library, **20**, 132, 136, 214
List scheduling, **98**–99
Loop, 78, 96
  data dependant loop, **81**, 108–109
  finite loop, **80**
  infinite loop, **79**–80
  loop folding, **78**, 80
  loop unrolling, **78**, 80

M

Mixed abstraction levels, **4**
Modeling for reuse, **147**
Modular decomposition, 132, 135, **141**, 143
Modular design, **133**–134, **141**–143, 217–248
Modularity, **18**–20, **131**–134, 141–143, 217–248
Multiplexer, **46**

O

Object Oriented VHDL, **150**
Operative part, **33**–34
See Datapath

P

Parallel transfer, **51**
Partitioning, **131**, 143, 205, 222, 240
Path probability, **109**
Physical level, **2**
PID, 143, **203**–205, 214
Pipeline Path-based scheduling, **115**–118

Pipelining, 12
  component pipelining, 43
  control pipelining, 37
  datapath pipelining, 52
  pipeline latency, 52
  pipeline stage, 52
PPS, 115
See Pipeline Path-based scheduling
Procedure calls, 82, **146**–149
Programmable architecture, **63**, 118

R

Register Transfer Level, **3**, 5, 11
Regularity, **18**
Reuse, 17–18, **127**–134, 155, 206–210, 227–240
  architecture, **133**
  behavioral component, **137**
  behavioral level, **135**–147, 203
  case study, **141**–144, 217–248
  component reuse, 18, **136**
  hierarchical, **133**
  process reuse, **18**
  Register Transfer Level, **135**
  VHDL modeling, **144**
RTL, **3**
See Register Transfer Level

S

Scheduled Path, **106**–107
Scheduling, 12–13, 16, 91, **94**
  Control Flow Based, **94**, 105–118
  Data Flow Based, **94**–96, 96–105
  execution thread, **91**–92
  I/O operations, **93**
  loop, **78**–81, 96
  programmable architecture, **118**–124
Sequencing graph, **50**
Specification, **204**, 218

State transition probability, **109**
Storage unit, 35, **45**
  constant register, **45**
  status register, **45**
  variable register, **45**
Structured design methodology, 18–20, 127, **130**–134, 141–143
SU, **45**
See Storage Unit
Super state mode, **92**
Switch, **46**
Synchronization model, **35**–39
System design, **5**, 240, 243
System level, **4**

T

Throughput, **44**, 53
Time and Resource Constrained scheduling, **102**
Timing unit, **2**–4
Transfer, **49**, 55–56
TRCS, **103**
See Time and Resource Constrained Scheduling

U

Unconstrained scheduling, **97**

V

VHDL, 11, **68**
  architecture, **68**
  behavioral description, **70**, 81–82
  conditional statements, **71**–72, 105
  interpretation modes, **93**
  iterative statements, **76**–78
  synchronization statements, **74**–75
  entity, **68**
  functions, **81**–82

Object Oriented VHDL,
    150–153
procedures, **81**–82
process, **68**–70, 83
sequential statements
  case, **71**
  if, **71**
  loop, **76**–78
  wait, **74**–75
signal, **83**–84, 87–88
types
  access, **89**
  array, **88**–89
  file, **89**
  record, **88**
  scalar, **88**
variable, **83**–84, 87–88

W

Window searching algorithm,
    217–248